POST 62

COOKBOOK

By Diane Freaney, Activist Author

ISBN: 978-1-962346-10-8 (e-book)
ISBN: 978-1-962346-11-5 (hardcover)
ISBN: 978-1-962346-09-2 (paperback)

US Copyright Registration Number: ####

Author photo by: Carson's Tavern
Editor and book designer: Kali Browne
Illustrator and cover designer: Matt Hosafros
Associate Editor: Edith Mahaffey

First Edition Published by Martin Pearl Press LLC

Some parts have been fictionalized in varying degrees, for various purposes.

For more information:

dianeyfreaney@martinpearlpresss.com
www.martinpearlpress.com

Martin Pearl Press LLC
2336 SE Ocean Blvd #230
Stuart, FL 34996-3310

Other Books by Diane Freaney

Coffee Table Books

The Art of Jimmy Tsutomu Mirikitani Exhibition Catalogue: A Traveling Exhibit of the Wing Luke Museum (2022). Emerson House Press.

Emerson Street Community, A Neighborhood Co-operative (2023). Martin Pearl Press LLC.

Living Community Challenge Series

Emerson Street Community, A Neighborhood Cooperative. (2023). Emerson House Press.

What Is This Place? MLK Jr. Membership Model. (2022). Emerson House Press.

What is This Place? Conversations Around Emerson Street House. (2021). Emerson House Press.

What is This Place? A Realtor's Tour of the Emerson Street House. (2022). Emerson House Press.

Complexity = Corruption Series

Complexity = Corruption / The US Tax System. (2023). Martin Pearl Press LLC.

Complexity = Corruption / Big Tech 5 / Amazon, Apple, Facebook, Goggle, Microsoft. (2022). Emerson House Press.

Complexity = Corruption / Big Tech 5 / Amazon. (2021). Emerson House Press.

Complexity = Corruption / Big Tech 5 / Apple. (2021). Emerson House Press.

Complexity = Corruption / Big Tech 5 / Facebook. (2021). Emerson House Press.

Complexity = Corruption / Big Tech 5 / Google. (2021). Emerson House Press.

Complexity = Corruption / Big Tech 5 / Microsoft. (2021). Emerson House Press.

Politics of Place Series

Surfside Condo Tragedy. (2022). Emerson House Press.

Living on the Medical Mile in Stuart Florida. (2022). Emerson House Press.

Higher Education Series

He Who Has the Gold Makes the Rules. (2021). Emerson House Press.

A New Model for the Creative Use of College Endowments to Reduce College Tuition. (2021). Emerson House Press.

A Systems Thinking Model to Grow Local Living Economies at For The Children Inc. (2022). Emerson House Press.

Preface

Post Cookbook

A simple idea – create a cookbook that veterans' organizations can sell to raise money for veterans and military families. It turns out that nothing is simple when the military is involved.

I have been challenged to dig deep to understand the history that binds military folks together. This is my attempt to summarize all that I have learned.

American Legion

"

A group of 20 officers who served in the American Expeditionary Forces in World War I were asked to suggest ideas on how to improve troop morale. One officer, Lt. Col. Theodore Roosevelt Jr., proposed an organization of veterans, which we know today as The American Legion.

"The original purpose of The Legion was to "preserve the memories and incidence of our association in the great war," helping those who had served in foreign wars to reintegrate into their hometowns while still remaining connected to those with whom they had served abroad.

"The Legion served as a support group, a social club and an extended family for former servicemen. After two planning caucuses held by a committee of officers who had the confidence and respect of their military comrades, they designed a constitution to govern the group and set up headquarters in New York City to begin work on its programs of relief, employment and Americanism.

"

American Legion Auxiliary

After the formation of The American Legion, a number of women's organizations wanted to become the official affiliation of The American Legion. The women who had served so faithfully during the trying days of the war wanted to continue to serve.

After careful consideration, the committee agreed that a new organization should be made up of the women most closely associated with the men of the Legion, and that these women would serve with the Legion, in peace as they had in war.

The committee decided to build a new organization from the ground up, so the Auxiliary could then carry forward the phases of Legion activities more suitably performed by women.

Sons of the American Legion

Founded in 1932, the Sons of The American Legion exists to honor the service and sacrifice of Legionnaires.

SAL members include males of all ages whose parents or grandparents served in the U.S. military and were eligible for American Legion membership. Members of The American Legion, American Legion Auxiliary and Sons of The American Legion comprise the Legion Family, which has a combined membership of nearly 3 million.

Although SAL has its own membership, the organization is not a separate entity. Rather, SAL is a program of The American Legion. Many Legionnaires hold dual membership in SAL.

American Legion Riders

American Legion Riders chapters are well known for their charitable work, which has raised hundreds of thousands of dollars for local children's hospitals, schools, veterans homes, severely wounded service members and scholarships.

Since 2006, Riders nationwide have participated in The American Legion Legacy Run to annually raise money for the Legacy Scholarship Fund, established to provide scholarships to children of U.S. military personnel killed since Sept. 11, 2001.

Military Funerals

A military funeral in the United States is a memorial or burial rite conducted by the United States Armed Forces for a Soldier, Marine, Sailor, Airman, Guardian or Coast Guardsman who died in battle, a veteran, or other prominent military figures or a president. A military funeral may feature guards of honor, the firing of volley shots as a salute, drumming and other military elements, with a flag draping over the coffin.

In the United States, the United States Army Military District of Washington (MDW) is responsible for providing military funerals. "Honoring Those Who Served" is the title of the program for instituting a dignified military funeral with full honors to the nation's veterans.

—https://en.wikipedia.org/wiki/Military_funerals_in_the_United_States

Road to Victory Military Museum

The Road to Victory Military Museum (https://www.facebook.com/roadtovictory/) provides a living history lesson to Martin County school children and adults who want to learn more about the history of the United States military.

The Stuart Air Show – November 8 - 10, 2024

The Stuart Air Show (http://stuartairshow.com) is right in my backyard at Whitham Field.

Whitham Field

Today, Witham Field with its three runways, serves most of the general aviation needs for Martin County and surrounding areas. The airport, just a few miles southeast of the City of Stuart, can accommodate most general aviation aircraft including smaller business jets, and is home to more than 200 private and business aircraft. It also plays host to the largest community event on the Treasure Coast, the annual Stuart Air Show. The airports fixed-based-operators (FBO's) act as the local terminals for all its users and those visiting Martin County from other locales.

The history of Witham Field is emblematic of the people and County in which it resides. Its very beginning is deeply rooted in duty, patriotism and generosity from the residents of Martin County. As with many rural airports in the 30's, Martin County Airport was just a grass landing strip. But with the rumblings of war overseas in the 40's, a group of visionary and patriotic private landowners convened and offered their property to the county, to serve their county."

—https://www.stuartairshow.com/team

Missing Man Table

A missing man table, also known as a fallen comrade table is a ceremony and memorial that is set up in military dining facilities of the United States armed forces and during official dining functions, in honor of fallen, missing, or imprisoned military service members. The table serves as the focal point of ceremonial remembrance, originally growing out of US concern of the Vietnam War POW/MIA issue (for more information, see: https://en.wikipedia.org/wiki/Vietnam_War_POW/MIA_issue).

The missing man table may be a permanent display in dining facilities, and is also traditionally part of military dining-in ceremonies and service balls. The ceremony may also be performed in conjunction with Veterans Day and Memorial Day services.

The missing man table consists of the following elements:

A small table set for one, symbolizing the isolation of the absent service member. The table is usually set close to, or within sight of, the entrance to the dining room. For large events, the missing man table may be set for six places representing each of the five armed services (Army, Navy, Marine Corps, Air Force, and Coast Guard), with the sixth symbolizing the civilians who died during armed conflict.

The table is round to represent the everlasting concern the survivors have for the missing.

One or more head covers may be placed upon the table to represent the armed service of the missing persons.

A white tablecloth to symbolize the pure intentions of the service members who responded to the country's call to arms.

A single rose in the vase symbolizing the blood that service members have shed in sacrifice to ensure the freedom of the United States of America. This rose also represents the family and friends who keep the faith while awaiting the return of the missing service members.

The red ribbon represents a love of country that inspired the service members to serve the country.

A slice of lemon on the bread plate that represents the bitter fate of the missing.

Salt sprinkled on the bread plate that symbolizes the tears shed by waiting families.

An inverted glass to represent the fact that the missing and fallen cannot partake.

A lit candle symbolizes a light of hope that lives in hearts to illuminate the missing's way home.

An empty chair to represent the absence of the missing and fallen.

Coming Home by Edith Mahaffey

“The real adjustment period was coming home,” Traylor said. “Ever since doing all these deployments, I’ve been home for a little over 5 years. Nothing has been the same for me. Every day is a real struggle.”

Being reintroduced into American culture, Traylor said that for some of his fellow soldier’s things like not being able to carry a weapon feels foreign and uncomfortable.

“Being able to go outside without a weapon is an adjustment because you get so used to that comfort of weapons and now you don’t have to have that anymore, “Traylor said. “Walking around large crowds is a little nerve-wracking. You have to tell yourself, ‘It’s OK, I’m in America. There’s not going to be any issues here today.’

“You avoid crowds and watch people. You’re always looking for opportunities of an attack that aren’t there anymore. It’s an adjustment that you don’t have to be on constant overlook anymore, but it is hard to comprehend.”

The biggest transition for Traylor is “loud noises”. “I was in a kill zone once. There was a mortar explosion about 50 feet behind me. I was on PT in my running gear when it exploded. My hearing was just gone for a few minutes, felt like longer. Smoke was everywhere. I never ran so fast in my whole life. I am just lucky that nothing hit me. Therefore, loud noises make me very nervous still. “

One more transition for Traylor is the adjustment of driving habits. When driving overseas, seeing something in a ditch, with the concern of an IED ever-present, is cause to swerve into the center of the road, a practice not accepted in America.

“You really have to convince yourself things are going to be OK, especially when driving,” Traylor said. “When you see trash on the side of the road, you realize it isn’t an IED so you can drive down the lane without going into the middle of traffic. I used to be a driver in Iraq and when I got back from that, I used to go in the middle of the road and get in trouble because I saw a trash bag or something on the side of the road, which over there is a big deal.

"That was a big thing to overcome. You can't be an aggressive driver like you were over there."

In addition to realizing there's no need to fear threats that were ever present while serving, Traylor said readjusting to life in general —which feels as though it's "passed you by" — can be a challenge in itself.

"You feel like everything has changed around you, like everything's different," Traylor said. "New movies have come out while you were gone, life just went on. It's an adjustment to get back into the swing of things.

"One problem a lot of us have is we get these high expectations of coming home, so it's a little disappointing when you get back to your life and realize there's still bills to pay.

"I've also noticed it's a lot easier over there because you have so much structure and know what's going on each and every day. Now you have to plan everything yourself. Over there you knew who you were going to see every day. You have to get used to making those plans. Housing was never an issue over there, but here I am still struggling to find a home for my family."

"How's your sleep?" I ask him.

"I don't," he answers.

Depending on the war, post-traumatic stress can have many expressions, but this war, because of its omnipresent suicide bombers and roadside explosives, seems to have disproportionately rendered its soldiers afraid of two things: driving and crowds.

Movie theaters, densely packed spaces—all can pose problems for soldiers, because marketplaces are frequent targets for explosions; so can any vehicle, because IEDs are this war's lethal booby trap of choice. Traylor manages his driving anxieties by leaving his early in the a.m., when there's no risk of traffic (especially under bridges, which militants in Iraq are always blowing up), and avoiding the right lane (in Iraq, one generally drives in the middle of the road to avoid setting off IEDs).

Once he gets to his destination, Traylor parks around the corner away from the crowded area and has managed to arrange his life so that he never encounters more than a handful of people. It is a real challenge going out to eat, which he handles by ordering in, taking carry out, or skipping entirely. I ask if he goes to restaurants in the off-hours. "Not very much," he answers.

Spend five minutes in Traylor's company, and it's hard not to be moved by the withdrawn contours of his life. He's in pain and can't sleep; but it's also hard not to marvel at his resilience. He's proud and uncomplaining; but he is comfortable talking about his struggles. When confronted with the reality and the struggle to find a permanent career—Traylor took several technical courses, and is still enrolled in a business management course, but still has not found his path.

If the Army's Medical Review Board no longer found him fit for duty, he wasn't going to protest. "You can't spend the rest of your life in the Army, just trying to heal," he says. "You're going to spend the rest of your life healing one way or the other anyway no matter where you do it."

In other words, nearly as many soldiers are dying at home today as are dying abroad. A soldier, on average, has moved from home, family, and friends and resided in two or more states; has traveled the world (deployed); been promoted four times; bought a car and wrecked it; married and had children; has had relationship and financial problems; and seen death, and came back to even more hardships than they faced abroad.

Written by Edith Mahaffey, Mother of Traylor

1 American flag crocheted by Edith Mahaffey and raffled off to raise funds for Chloe, her 4 year old niece going through chemotherapy.

Dedicated to Bruce Hudson

Home Going Celebration

Bruce Dennis Hudson

June 21, 1947 - January 15, 2024

Saturday, February 17th, 2024, at 11:00 am
American Legion Post 126
3195 NE Savannah Rd
Jensen Beach, Florida 34957

Order of Service

Moderator..........Tom Scullin

Invocation..........Chaplain Mike Foley

Reflections..........Paul Hudson

Poem Reading..........Krissy Faulk

Bagpipe "Amazing Grace"..........Hobe Sound VFW Honor Guard

Presentation of the US Flag..........United States Marine Corps

Roll Call..........Hobe Sound VFW Honor Guard

Marine Corps Hymn.......... Hobe Sound VFW Honor Guard

Rifle Salute......... Marine Corps League of PSL, and Hobe Sound VFW Honor Guard

TAPS..........Hobe Sound VFW Honor Guard

*******RECEPTION TO FOLLOW ON THE BACK PATIO*******

Acknowledgement

The family of Bruce Hudson extends its most gracious appreciation and thanks to
The American Legion Post 126, Hobe Sound VFW Honor Guard,
Marine Corps Legion of Port St Lucie, The United State Marine Corps,
and all relatives and friends.

We are so grateful for your many acts of kindness
and support during our bereavement.

Bruce Dennis Hudson
June 21, 1947 - January 15, 2024

Bruce Dennis Hudson passed away peacefully surrounded by his loving family and friends on January 15th, 2024.

Bruce was born on June 21st, 1947, in Camden, New Jersey. As a very young man, he bravely joined the United States Marine Corps where he proudly served in Vietnam from 1965 to 1967. His time there was quite scenic, filled with monsoons, rice paddies and giant snakes. Though he was wounded in the leg and arm, in true Bruce fashion, he was able patch himself up and get his butt out of there.

When he got back to the good ole USA, Bruce moved to Florida, where his family had moved to while he was overseas. There he joined the Union 60 Heat and Frost Asbestos Workers, where he and his brothers became lifelong friends. He was extremely hard working and dependable.

Bruce spent his free time helping other Veterans and rarely missed a Memorial Day event, Veterans Day event, or any other opportunity to show his support. He helped line up and march the Veterans for 30 years at the parades and served as the Commander of the Amvets Post 92 at least 4 times. He also packed and mailed hundreds of care packages for the troops overseas. His dedication and love for Veterans and his Community was unsurpassed.

Bruce is survived by his wife Debra Hudson of 46 years. He is also survived by their daughter Ashley Sears, her husband James Sears and their sons Carson, Gabriel, and Iain. Bruce and Debra helped raise their grandsons after their daughter Ashley had a stroke at the age of 36. They also helped raise an extended daughter, Krissy Faulk from a young age. He was a loving family man who never needed to discipline any of the kids. They loved and respected their POP immensely.

Bruce had three brothers, Paul (Gary and Jason Hudson who have also passed and are deeply missed). Bruce also had two sisters, Christa Ragland, Elizabeth Gordon and two sisters-in-law Debbie Hudson and Kim Hudson who through the years were as close as sisters. He also loved all his nieces and nephews and looked forward to family gatherings with them.

You all know Bruce would love to hear a loud HOORAH from Heaven's gate, so lets give him one!

SEMPER FI, DO OR DIE! IMPROVISE, ADAPT, AND OVERCOME!

The Marines Prayer

Almighty Father, whose command is over all and whose love never fails, make me aware of Thy presence and obedient to Thy will. Keep me true to my best self, guarding me against dishonesty in purpose and deed and helping me to live so that I can face my fellow Marines, my loved ones and Thee without shame or fear. Protect my family. Give me the will to do the work of a Marine and to accept my share of responsibilities with vigor and enthusiasm. Grant me the courage to be proficient in my daily performance. Keep me loyal and faithful to my superiors and to the duties my country and the Marine Corps have entrusted to me. Make me considerate of those committed to my leadership. Help me to wear my uniform with dignity, and let it remind me daily of the traditions which I must uphold.

If I am inclined to doubt, steady my faith; if I am tempted, make me strong to resist; if I should miss the mark, give me courage to try again.

Guide me with the light of truth and grant me wisdom by which I may understand the answer to my prayer.
AMEN.

Contributors

Rebecca Baruffi
Denise Bonadeo
Cangianellis Family
Michon Chambon
Livia Coates
Sandy Coughlin
Vivian Currell
Michelle Divine
Chris Eisenberg
Denise Gauthier
Lollie Higbie
Barb Hoertz
Lucille Lambert
Will Maiden
Cheryl McClure
Nancy McCool
Terry Porter
Shirley Rearick
Margo Roney
Sherry Sanderson
Terry Sanderson
Freda Smith
Ella Swinarski
Joanie Thomas
Joane Valencia
Kendall White

Table of Contents

BREAKFAST & BRUNCH

BACON CANDY STICKS

Contributed by Viv Currell

2 lbs. bacon strips, twisted
1 Tbsp. brown sugar
1 tsp. cayenne

Carefully twist bacon strips individually to form long curled twists. Place on shallow baking pan, each strip touching each other. Mix brown sugar and cayenne together and sprinkle over each bacon twist. Bake 40 minutes at 400° turning once.

TATER TOT BREAKFAST CASSEROLE

Contributed by: Joanne Valencia

1 frozen bag of tater tots
12 eggs
2 c. shredded cheese
½ lb. bacon
1 tsp. onion powder
½ tsp. garlic powder
Salt and pepper to taste
Sliced green onions

Preheat oven to 350° F.

Spray a 9 x 13-inch baking dish.

Place tater tots in a single layer on the bottom of the baking dish.

In a medium bowl whisk together eggs until smooth. Stir in the onion powder, garlic powder, salt and pepper. Mix in crumbled bacon. Then, stir in 1 cup shredded cheese.

Pour egg mixture over the tater tots. Make sure the egg is spread out all on the bottom. Top with the remaining cheese.

Bake uncovered for about 35-40 minutes. The egg mixture should be fully cooked and no longer jiggly.

Yield: 6 *Prep Time: >15 minutes* *Total Time: >1 hour 15 minutes*

SAUSAGE HASHBROWN BREAKFAST CASSEROLE

Contributed by: Joanne Valencia

1 lb. frozen bag of hashbrowns
1 lb. pork breakfast sausage
2 c. shredded cheese
½ onion finely chopped
1 tsp. salt
¼ tsp. pepper

Preheat oven to 375° F.

Spray a 9 x 13-inch baking dish.

In a sauté pan cook onions and brown and crumble sausage. Drain excess grease. If you would like, you can take out the bacon and brown the hashbrowns a bit with the grease to give it flavor.

In a medium bowl whisk together eggs, milk, 1 cup of cheese, and salt and pepper.

In the baking dish spread the hashbrowns. Next pour the egg mixture on top of that. Sprinkle with the remaining cheese.

Pour egg mixture over the tater tots. Make sure the egg is spread out all on the bottom. Top with the remaining cheese.

Bake covered with aluminum foil and for about 30 minutes. Take off the foil and bake for about 10 minutes longer. Keep an eye to make sure it does not brown too much.

Yield: 6   *Prep Time: >15 minutes*   *Total Time: >1 hour 15 minutes*

OVERNIGHT CROCK POT OATMEAL

Contributed by Viv Currell

Place two sliced apples, ¼ cup brown sugar, 1 tsp. cinnamon, and a pinch of salt in the bottom of a crock pot. Pour in 2 cups of oatmeal, 2 cups of milk and 2 cups of water. DO NOT STIR! Cook overnight 8-9 hours on low.

MAPLE CREAM CHEESE FRENCH TOAST CASSEROLE

Contributed by Beverly Spiegel

10 c. slightly stale bread (brioche is good)

8 oz. cream cheese

3-4 Tbsp. sugar (may use both white and brown)

8 large eggs, slightly beaten

1 ½ c. whole milk

⅔ c. half and half

½ c. maple syrup

2 Tbsp. vanilla

Powdered sugar

Place cubed bread into a buttered 13 x 9-inch pan. In separate bowl, beat together cream cheese and sugars till well incorporated. Add eggs a little at a time, mixing well. Add milk, half and half, maple syrup and vanilla and mix well. Pour cream cheese mixture over bread and gently stir with a fork or spoon till all bread is evenly coated. Cover and refrigerate 8 to 24 hours. When ready to bake, set oven at 375°. Remove dish from fridge and let sit for 30 minutes. Bake uncovered 45-50 minutes or until golden brown. Sprinkle with powdered sugar and serve with warm maple syrup.

APPETIZERS

COLD VEGGIE PIZZA

Contributed by Rebecca Baruffi

1 can crescent rolls
4 oz. cream cheese softened
½ cup mayonnaise
½ package Hidden Valley ranch mix
1 ½ cups mixed cut up vegetables, chopped in small dice. (peppers, carrots, celery, broccoli, radishes, zucchini, cauliflower). Mix up colors.
⅓cup shredded mozzarella cheese

Spread out crescent rolls, press thin and attach seams. Bake as directed. 375° F 8-10 minutes.

Mix cream cheese, mayonnaise and ranch dressing until smooth spread over cooled crust. Serve chilled.

Sprinkle finely chopped veggies over top. Sprinkle mozzarella over vegetables. Cut with pizza cutter into squares.

Yield: 12 *Prep Time: <15 minutes* *Total Time: 20 minutes*

BUFFALO CHICKEN DIP

Contributed by Nancy McCool

½ cup blue cheese dressing
1 pkg. cream cheese, diced
1 pkg. Purdue chicken boiled and diced
½ cup Pete's hot sauce
3 cheeses shredded on top

Layer cream cheese, blue cheese and chicken. Sprinkle with 3 cheese mixture on top.

Bake for 25-30 minutes in a preheated 350° F oven.

Serve with celery, carrots and assorted crackers.

Yield: 6 *Prep Time: >10 minutes* *Total Time: 1 hours*

PARMESAN CHEESE BITES

Contributed by Chris Eisenberg

1 c. all-purpose flour
¼ tsp. ground red pepper
⅔ c. grated parmesan cheese
½ c. butter or margarine
2 Tbsp. milk

Preheat oven to 350°. Pulse first 4 ingredients in food processor till blended. Shape dough into 2 to 4-inch logs. Wrap in plastic wrap and place in airtight container to chill for 8 hours (see note). Once chilled, cut dough into ¼-inch slices and place on lightly greased baking sheet. Brush with milk. Bake at 350° for 12-15 minutes till lightly browned. Yield: 2 ½ dozen.

Note: Dough may be frozen up to 3 weeks and thawed in refrigerator overnight if desired. Also note: Baked cheese bites can be frozen up to 3 weeks. Thaw at room temperature 2 hours before serving.

DRUNKEN HOT DOGS

Contributed by Chris Eisenberg

6 pkgs. red hot dogs
1 bottle ketchup
1 bottle bourbon

Put bourbon and ketchup in chafing dish and heat till very hot. Cut hot dogs into bite size pieces, and boil in water. Drain and place hot dogs in sauce. Use colored toothpicks to celebrate any patriotic holiday.

Yield: 8-10 Prep Time: >5 minutes Total Time: 3 hours

MARINATED CHEESE

Contributed by Viv Currell

½ c. evoo
½ c. white wine vinegar
1 jar (2-oz.) diced pimento, drained
3 Tbsp. fresh chopped parsley
3 Tbsp. minced green onion
3 cloves garlic, minced
1 tsp. sugar
¾ tsp. dried basil
½ tsp. salt
½ tsp. ground black pepper
8-oz. pkg cheddar cheese
8-oz. pkg cream cheese (cold)

In a jar with tight fitting lid combine extra virgin olive oil, wine vinegar, pimento, parsley, green onion, garlic, sugar, basil, salt and pepper. Shake vigorously and set aside. Cut block of cheddar in half lengthwise and then cut crosswise to ¼-inch slices. Repeat with cream cheese. Arrange cheddar and cream cheese in alternating layers into a shallow baking dish, with slices standing on edge. To serve, remove from baking dish and keeping the layers, place on serving platter. Spoon extra marinade over cheese. Serve with assorted crackers and crusty bread.

PEPPERONI DIP

Contributed by Nancy McCool

½ lb. pepperoni
2 blocks of cream cheese softened
1 frozen block of spinach
1 cup shredded cheese

Mix all the ingredients and stuff pepperoni mixture into individual prepared muffin pan.

Bake in preheated 350° F oven for 25-30 minutes

Yield: 10-12 *Prep Time: >10 minutes* *Total Time: 40 minutes*

QUICK AND EASY CREAM CHEESE APPETIZER

Contributed by Viv Currell

Place a softened block of cream cheese on serving dish. Cover with warmed pepper jelly and cooked crumbled bacon pieces. Serve with crackers, bread sticks or crusty bread.

SWEET POTATO SKEWERS

Contributed by Viv Currell

2 medium sweet potatoes, peeled and cut into 1-inch cubes

12-oz. precooked chicken andouille sausage (about 4 links) or 12-oz. precooked kielbasa

1 can pineapple chunks, drained

2 ½ tsp. evoo, divided

1 tsp. brown sugar

½ tsp. kosher salt

¼ tsp. black pepper

1 Tbsp. butter

¼ c. plain Greek yogurt

¼ c. Dijon mustard

Preheat oven to 425°. Line a large baking sheet with parchment paper; set aside. Place sweet potato chunks in a large bowl; drizzle with 1 ½ teaspoon extra virgin olive oil and sprinkle with brown sugar, salt and pepper. Toss to evenly coat. Arrange in a single layer on baking sheet and bake 20-25 minutes, turning them once half-way though baking time. While potatoes bake, mix together yogurt and Dijon in a small bowl for a dipping sauce. Set aside. Heat the remaining teaspoon of extra virgin olive oil in a large skillet over medium heat. Cut sausages into 1-inch pieces and arrange sausage pieces cut sides down in a single layer into the heated skillet. Sear on cut side of each piece, crisping the exterior; turn and crisp the other side. Remove sausages from skillet; add 1 tablespoon butter to the drippings; place drained pineapple chunks into skillet and sear for 1 minute. Assemble skewers by poking a toothpick thru the center of a pineapple chunk, then a sweet potato cube and then a sausage piece. Transfer to serving platter; serve skewers warm or room temperature with dipping sauce.

MEATBALL YUMMIES

Contributed by Chris Eisenberg

1 lb. ground beef
⅓ c. dry bread crumbs
1 egg
2 Tbsp. Prime Choice steak sauce
2 Tbsp. oil
⅓ c. Prime Choice steak sauce
2 Tbsp. light brown sugar
2 Tbsp. butter or margarine

Combine beef, bread crumbs, egg and 2 tablespoons of Prime Choice steak sauce. Mix and shape into 1-inch balls. Brown in oil. Drain fat from skillet. Combine ⅓ cup Prime Choice, brown sugar and butter with meatballs in skillet. Simmer, covered for 15 minutes; until done. Makes 2 dozen meatballs.

VIV'S TOMATO SALSA

Contributed by Viv Currell

¼ c. chopped Roma tomatoes
½ chopped red bell pepper
½ chopped green bell pepper
¼ c. chopped red onion (optional)
3-4 chopped green onions
4 tsp. chopped fresh jalapeno, seeded
2 large garlic cloves, minced
½ c. chopped fresh cilantro
2 Tbsp. lime juice
½ tsp. cumin
½ tsp. kosher salt
½ tsp. black pepper
Drizzle of evoo, optional
Tortilla chips

Stir together all ingredients; drizzle lightly with extra virgin olive oil if desired; chill and serve with tortilla chips.

SWEET AND SOUR MEATBALLS

Contributed by Chris Eisenberg for Cheryl McClure

3 ½ lbs. ground meat
3 ½ lbs. sausage
7 eggs
2 ½ c. soft bread crumbs
¾ c. soy sauce
5 c. ketchup
2 c. packed brown sugar
¾ c. white wine vinegar

Mix together meat, sausage, eggs and bread crumbs. Form into balls and sauté in frying pan until browned on all sides. Drain. Combine the rest of the ingredients and cook on low heat. Simmer 30 minutes, stirring occasionally. Can be frozen for later; freeze sauce and meatballs together. Reheat in 350° oven for 20 minutes to serve.

OYSTERS ROCKEFELLER

Contributed by Chris Eisenberg for Will Maiden

2 sticks butter
1 pkg. frozen chopped spinach
3 medium onions, grated
6 stalks celery, grated
5 Tbsp. dried parsley flakes
4 cloves crushed garlic
6 Tbsp. crushed bread crumbs
1 Tbsp. Pernod
Healthy dash of cayenne, salt/pepper
2 doz. oysters, shucked but with shells

Melt butter in large saucepan. Add all ingredients except oysters. Mix and let bubble gently 30 minutes. Dry oysters and place in shells on a bed of rock salt in aluminum pan. Top with sauce; bake 10 minutes at 350°. Then turn on broiler and leave until sauce is bubbly. Let cool one minute and serve with a small cup (2-oz.) of butter for each person.

BAJA ROLL-UPS

Contributed by Viv Currell

½ c. shredded Mexican blend cheese
1 package taco seasoning
2-3 green onions, finely chopped
1 – 2 ¼ oz. can sliced black olives, drained
1 8-oz. package cream cheese
2 oz. chopped green olives
5 – 10-inch flour tortillas

Stir in seasonings, green onions and green olives. Spread thinly on flour tortilla; then layer black olives on top of cream cheese mixture. Roll up tortillas, wrap in plastic wrap and chill for at least an hour. Just before serving, slice into pinwheels, arrange on serving platter and serve with salsa or salsa verde.

Variation: Mix spices of choice with softened cream cheese. Spices can be fajita seasoning, fresh or dried cilantro, chili powder, cumin, garlic powder, adobo or two packets of taco seasoning.

BAJA ROLL-UPS DE LUJO (DELUXE)

Contributed by Viv Currell

½ c. shredded Mexican blend cheese
1 – 8-oz. pkg. cream cheese
1 oz. (up to 4 Tbsp) dry spices of choice (fajita seasoning, dried cilantro, chili powder, cumin, garlic powder, adobo or 2 packets of taco seasoning)
4 green onions, chopped
2 Tbsp. chopped seeded fresh jalapeno
15 oz. can drained black beans
11 oz. can drained crispy sweet corn
1 – 2 ¼ oz. can drained black olives
½ c. crushed green olives with pimento
2 Tbsp. chopped fresh cilantro
10-inch flour tortillas

Whip cream cheese till soft and spreadable. Stir in seasonings, mix well. Fold in rest of ingredients except tortillas. Spread cream cheese mixture on flour tortillas. Roll up, wrap in plastic wrap and chill at least one hour. Before serving, slice into pinwheels, arrange on serving platter and serve with salsa, salsa verde or sour cream.

SOUPS

POTATO SOUP

Contributed by Vivian Currell

1 lb. diced potatoes
¼ c. chopped onion
1 ¼ tsp. salt
2 c. boiling water
¼ c. butter or margarine
¼ c. flour
2 c. milk
2 c. grated sharp cheddar cheese
1 Tbsp. Worcestershire
cooked bacon and chopped chives

Boil potatoes, salt and onion in water until potatoes are done. DO NOT DRAIN! Melt butter in a saucepan; add flour and cook briefly, stirring continuously. Gradually add milk and cook to thicken. Remove from heat. Stir in cheese to melt; then add potato mixture and Worcestershire. Heat to serving temperature but do not boil. Sprinkle with chopped cooked bacon pieces and chopped chives before serving.

HERBED GAZPACHO

Contributed by Chris Eisenberg for Will Maiden

2 (14 ½ oz.) cans peeled tomatoes (with liquid)
½ lb. unpeeled cucumber, finely chopped (1 ½ c.)
½ c. finely chopped scallions
12 c. finely diced celery
¼ c. chopped fresh basil
2 Tbsp. herbed or red wine vinegar
1 small garlic clove, minced
½ tsp. black pepper/pinch of salt

Combine tomatoes and their liquid with all remaining ingredients in a large bowl. Stil well to break up tomatoes. Cover the bowl and refrigerate the gazpacho 4 hours or until chilled. Stir the soup to reblend before serving and ladle it over ice cubes if desired. Makes 4 servings.

FRENCH ONION SOUP

Contributed by Chris Eisenberg

6 oz. melted butter
½ pt. dry sherry
1 ½ qt. chicken stock
3 lbs. Vidalia onions, peeled and thinly sliced
1 ½ qt. beef stock
Salt and pepper to taste
8 oz. pkg. mozzarella cheese
½ lb. Swiss cheese, grated
Croutons

Sauté onions. Add sherry to onions and simmer. Add beef and chicken stock. Bring to a boil and simmer on low for 30 minutes. Divide soup into individual crocks and add croutons. Top with grated cheeses. Broil till cheeses are lightly browned.

BRUNSWICK STEW

Contributed by Chris Eisenberg

1 large onion, chopped
6 chicken thighs (skinless)
1 large bag frozen mixed vegetables
3 potatoes, skinned and cut
Fresh tomatoes, cut in quarters
3 small cans V-8 juice
Salt and pepper
1 Tbsp. parsley, chopped
1 Tbsp. Creole spice
1 lb. fresh or frozen shrimp

Put all ingredients in a slow cooker or a large pot and let simmer all day or 4 hours in pot. Use as many fresh vegetables as possible. Serve with biscuits and homemade jam.

SALADS, SAUCES & DRESSINGS

ASIAN NOODLE SALAD

Contributed by: Denise Bonadeo

½ c. sliced almonds
8 Tbsp. sesame seeds
7 Tbsp. rice vinegar
4 Tbsp. sugar
1 tsp. salt
1 tsp. pepper
½ c. canola oil
1 16 oz. bag of coleslaw mix
2 pkgs. ramen noodles broken up (without the seasoning)

In a skillet over medium low to medium high heat, toast almonds and sesame seeds until light brown.

In a small bowl, combine vinegar, sugar, salt, and pepper, and whisk in oil.

In a large bowl combine coleslaw mix and ramen noodles. Add contents of small bowl and mix well. Toss in the almonds and seeds just before serving.

Yield: 6-8 *Prep Time: >10 minutes* *Total Time: 20 minutes*

OVERNIGHT SALAD

Contributed by Terry and Sherry Sanderson

1 large head cauliflower
1 head lettuce
1 lb. cooked bacon, crumbled
2 c. mayonnaise
1 c. parmesan cheese
1 tsp. lemon juice
2 dashes Worcestershire
Salt and pepper to taste

In a medium bowl, mix together mayonnaise, parmesan cheese, lemon juice, Worcestershire and salt and pepper. Break up lettuce in bottom of a large bowl. Break up cauliflower on top of lettuce. Sprinkle bacon on top of cauliflower. Spread mayo mixture on top; cover and refrigerate for 24 hours. Serve – stir up and put in a fancy bowl.

SHAVED CARROT SALAD

Contributed by Viv Currell

3 shaved carrots
2 Tbsp. sesame oil
1 tbs. soy sauce
1 tsp. maple syrup
2 Tbsp. chopped cilantro
Sea salt to taste
Juice of 1 lemon
Roasted sesame seeds

Add shaved carrots to serving bowl. Toss with all other ingredients. Chill and serve.

CURRIED CASHEW CHICKEN SALAD WITH DIJON DRESSING

Contributed by Viv Currell, Slightly Stolen from The Cangianellis

Salad

2 c. cooked, chopped chicken
¾ c. cashew nuts
¾ c. red grapes, halved
1 green onion, chopped
½ c. frozen peas
½ c. chopped celery

Dressing

1 c. mayonnaise
2 Tbsp. Dijon mustard
2 Tbsp. lemon juice
2 Tbsp. sugar
½ tsp. curry powder

Toss salad ingredients together in a medium bowl. Mix dressing ingredients in small bowl and whisk until well blended. Top each salad serving with dressing.

DILLED CHEESE AND SALAMI SALAD

Contributed by Viv Currell

½ lb. Swiss cheese
6 oz. thick sliced salami
1 thin-skinned cucumber
2 Tbsp. red wine vinegar
2 tsp. Dijon mustard
1 tsp. dried dill weed
1 small garlic clove, minced
½ tsp. salt
⅛ tsp. white pepper
2 Tbsp. evoo
¼ c. salad oil
Lettuce leaves

Cut Swiss cheese and salami in matchstick size strips. This should yield about 2 cups of cheese and 1 ½ cups of salami. Cut cucumber in half lengthwise then slice thinly. Place cheese, salami and cukes in large bowl. Cover and chill while preparing dressing. Mix vinegar, mustard, dill, garlic, salt and pepper with a fork in small bowl. Gradually beat in olive oil then salad oil until dressing is well blended and slightly thickened. Drizzle dressing over chilled cheese/salami/cuke mixture. Toss lightly and serve on lettuce leaves.

Yield: 4

BROCCOLI SALAD TO DIET FOR

Contributed by Chris Eisenberg for Lollie Higbie

3 c. raw broccoli florets
4 c. mozzarella cheese, shredded (large cut)
1 red onion, diced chunky
1 lb. fried bacon (optional)

Dressing Mix

½ c. white sugar
1 c. mayonnaise
2 Tbsp. white vinegar

Mix together dressing ingredients until smooth and add to broccoli with other ingredients above. Easier to mix in a third of the dressing at a time. Chill and serve.

BLT MACARONI SALAD

Contributed by: Joanne Valencia

1 16 oz. box elbow macaroni
1 ½ c. mayonnaise
¾ c. sour cream
3 Tbsp.. granulated sugar
2 Tbsp.. apple cider vinegar
1 ½ tsp. celery salt
1 ½ tsp. garlic powder
1 ½ tsp. smoked paprika
¼ tsp. dried dill
Ground pepper, to taste
5 green onions
4 Roma tomatoes
2 c. diced romaine lettuce
½ lb. bacon (cooked and crumbled)

Cook and drain pasta according to the directions. Drain well.

Prepare the dressing while pasta is cooking.

In a large bowl whisk mayonnaise, sour cream, sugar, vinegar, celery salt, onion powder, smoked paprika, dill, and pepper to taste.

Cut up the tomatoes, green onion, lettuce and bacon.

Add the cooked and drained pasta to the bowl with the dressing. Stir in the tomatoes and green onions. Mix evenly until distributed.

Cover with plastic wrap and pop it in the fridge to chill. Just before serving add the lettuce and bacon.

Mix well and serve.

Yield: 8 *Prep Time: >30 minutes* *Total Time: >35 minutes*

VEGETABLES

AWESOME BROCCOLI-CHEESE CASSEROLE

Contributed by: Unknown

1 (10.75 oz.) can condensed cream of mushroom soup

1 c. mayonnaise

1 egg beaten

¼ c. diced onion

3 (10 oz.) frozen packages broccoli

8 oz. shredded sharp cheddar cheese

Salt and pepper to taste

Paprika

Preheat oven to 350° F. Butter a 13 x 9-inch baking dish.

In a medium mixing bowl whisk together condensed soup, mayonnaise, egg and onion.

Place frozen broccoli into a large mixing bowl. Break up the broccoli. Using a rubber spatula scrape soup and mayonnaise mixture on top of the broccoli, mix well. Spread mixture into prepared baking dish and smooth top of casserole. Season vegetables with salt and pepper, and sprinkle with paprika.

Bake 350° for 45-55 minutes or until tender.

Yield: 8 *Prep Time: >10 minutes* *Total Time: 75 minutes*

LOADED BAKED POTATO CASSEROLE

Contributed by Kendall White

1 pkg. Bob Evans mashed potatoes (22 oz.)

1 container garlic and herb Alouette soft cheese (6.5 oz.)

½ c. milk

6 cooked and crumbled bacon strips

1 tsp. salt

1 c. shredded cheddar cheese

Mix all ingredients together over low heat. Pour into a baking dish. Bake at 350° for 25-30

BAKED CORN CASSEROLE

Contributed by Rebecca Baruffi

1 stick butter
1 can creamed corn
1 can whole kernel corn
4 oz. sour cream
2 eggs beaten
1 small box of Jiffy cornmeal mix

Melt butter and mix with eggs in a large bowl. Add remaining ingredients and mix well with a wooden spoon.

Spread in a 2 - 2 ½-quart prepared baking dish.

Bake at 350° F for 40-45 minutes.

Cake is done when a knife comes out clean.

Yield: 8-10 *Prep Time: >10 minutes* *Total Time: 50 minutes*

AU GRATIN PEAS AND POTATOES

Contributed by: Joanne Valincia

6 slices of bacon
1 medium onion, shopped
4 c. potatoes, peeled, sliced and cooked
½ tsp. salt
1 pkg. frozen peas, cooked and drained
2 c. sharp cheddar cheese, divided
½ c. mayonnaise
½ c. milk

In a large skillet, cook bacon until crisp. Remove with a slotted spoon to paper towels. Drain, reserving 1 tablespoon of the drippings. Sauté the onion until tender in the bacon drippings.

Layer the potatoes, salt, peas, 1 cup of cheese, and the bacon. Reduce heat and cover. Simmer until heated through, about 10 minutes.

Combine mayonnaise and milk until smooth; pour over bacon. Sprinkle with the remaining cheese. Remove from the heat and let stand for 5 minutes before serving.

Yield: 6 *Prep Time: >10 minutes* *Total Time: 30 minutes*

POTATOES AU GRATIN

Contributed by: Rebecca Baruffi

3 medium russet potatoes
1 c. heavy cream
1 tsp. kosher salt
½ tsp. pepper
1 clove garlic
2 oz. grated Swiss cheese

Peel and cut potatoes into ⅛-inch slices. In medium saucepan combine potatoes, heavy cream, salt and pepper. Heat just under boiling point.

Rub a medium baking dish with garlic and then spray lightly with vegetable spray. Layer potatoes evenly in prepared pan. Sprinkle cheese evenly over the surface. Bake until the potatoes are golden brown and tender.

Bake at 375° for 30-40 minutes or till tender. Remove from oven and let rest for 10 minutes.

Yield: 4 *Prep Time: >20 minutes* *Total Time: 50 minutes*

SPINACH AND RED PEPPER COUSCOUS

Contributed by Chris Eisenberg for Will Maiden

1 medium red bell pepper, diced
½ large red onion, diced
3 large garlic cloves, minced
2 Tbsp. olive oil
¼ c. Nakano seasoned rice vinegar-original
½ tsp. each-dried basil and oregano
¼ tsp. each salt and pepper
1 can currants or raisins
1 ½ c. uncooked couscous
1 ½ c. broth
2 c. fresh spinach leaves

In large saucepan, sauté bell pepper, onion and garlic in oil, stirring occasionally, till onion is soft (6 minutes). Add vinegar, herbs, salt and pepper; cook one minute. Stir in broth and currants; bring to a boil. Stir in couscous and spinach. Cover and remove from heat. Let stand 5 minutes. Makes 4-6 servings.

SOUTHERN POTATO CASSEROLE

Contributed by Chris Eisenberg

1 lb. bag shredded hash browns
1 large container fat-free sour cream
1 small onion, diced
2 cans cream of mushroom, celery or chicken soup
8 oz. shredded cheese
1 small jar salsa (optional)
cayenne pepper (optional)

Preheat oven to 350°. Mix all ingredients together (except shredded cheese) and place in a 9 x 13-inch baking dish. Top with shredded cheese. Bake for 45 minutes. Let set for 5-10 minutes and cut into squares for serving. May add salsa and/or sour cream as topping if desired.

minutes. During last 5 minutes of baking time, top with Durkee's crispy onions, crumbled.

BAKED ZUCCHINI

Contributed by Livia Coates

4 small zucchini
6 Tbsp. butter
¾ c. chopped carrots
½ c. chopped onion
2 ¼ c. herb flavored croutons
10 oz. can cream of chicken soup
½ c. sour cream

Preheat oven to 350. Cut zucchini in half lengthwise, slice crosswise in ½-inch pieces (3 cups). In a skillet on medium heat, melt 4 tablespoons butter; stir in zucchini, carrots, and onions. Cover pan and steam for 4-5 minutes. Stir in 1 ½ cups croutons. Remove skillet from heat. Stir soup and sour cream together and add to zucchini mixture. Spoon into 1 ½ quart baking dish. Crumble remaining croutons over top of zucchini mixture. Melt 2 tablespoons butter and drizzle on top. Bake for 30 minutes.

MEXICAN STUFFED PEPPERS

Contributed by Denise Bonadeo

6 poblano peppers
1 lb. ground beef
½ lb. ground chorizo
1 small onion
2 stalks of celery diced
Olive oil
1 can green chiles
1 can tomato sauce
1 package taco seasoning
Garlic, to taste
Salt and pepper to taste
4 oz. shredded Colby jack cheese
French's french-fried onions

Preheat oven at 350° F.

Blanch the peppers by adding them to boiling water and cooking for 5-7 minutes, remove from water, salt and let dry.

Sauté the onions and celery in heated oil for 3-5 minutes. Add the ground beef and chorizo, season with garlic, salt and pepper. Cook until brown. Drain fat and add green chiles, tomato sauce and taco seasoning, mix well.

Place green pepper halves in a 13 x 9-inch pan, fill each pepper with meat mixture. Top each with shredded cheese.

Put peppers in the oven and bake for 30-35 minutes. Remove from oven and sprinkle tops with fried onions. Place back into oven and continue baking until the cheese is melted and the onions are crisp.

Yield: 6 *Prep Time: >10 minutes* *Total Time: 1 hour*

COLLARD GREENS WITH HAM BONE

Contributed by: Joanne Valencia

2 tbs. vegetable oil
1 leftover hambone with meat still attached
64 oz. unsalted chicken stock
2 tbs. garlic powder
2 tbs. onion powder
2 lbs. chopped collard greens
1 tbs. seasoning salt
½ tsp. black pepper
¼ tsp. red pepper flakes
4 tbs. unsalted butter

Place a large stockpot or 6-quart Dutch oven over medium heat. Once hot, add the oil, place the ham bone and get a good sear on all sides, about 5 minutes.

Carefully pour in the chicken stock along with garlic powder and onion powder into the pot. Stir and combine. Bring pot to a boil, cover and lower to a simmer for at least 60 minutes. Using tongs take the ham bone out and set aside to cool. Add the collard greens to the pot. You need to do this in batches, or they will wilt. Add the seasoning salt, black pepper, and chili flakes. Stir to combine. Bring to a boil, cover, and reduce to a simmer for at least 2 hours or until desired tenderness. Let ham bone cool to touch and pull any remaining meat and add to the pot, stir in the butter and let it melt.

Serve. Leftovers freeze well.

Yield: 4 *Prep Time: >30 minutes* *Total Time: 4 hours 5 minutes*

MISCELLANEOUS

GOURMET BUTTERS

Contributed by Viv Currell

For all recipes, mix ingredient by chopping; placing on waxed paper, parchment, foil or plastic wrap. Roll up cylinder style, twisting ends. Make sure it's wrapped airtight. Store in fridge up to 2 weeks; freezer 3 months.

Porcini/Red Wine Butter (for steaks, chops, chicken): 1 stick unsalted butter at room temp; 2 tablespoons finely chopped dried porcini mushrooms, 1 tablespoons red wine; kosher salt

Chipotle-Lime Butter (pork, chicken, fish, shrimp): 1 stick unsalted butter at room temp; 1 canned, chopped chipotle chile in adobo; 1 teaspoon lime zest, kosher salt)

Blue Cheese Chive Butter (vegetables, potatoes, burgers, steak): ½ stick softened butter; 1/3 cup crumbled blue cheese; ½ cup chopped chives; cracked black pepper

Turmeric-Mustard Seed Butter (steamed rice): 1 stick unsalted butter, softened; 2 teaspoons dried turmeric, ¼ teaspoon brown mustard seeds, toasted; ½ teaspoon yellow mustard seeds, toasted; kosher salt

Bacon-Bourbon Butter (Waffles, pancakes, French toast): 1 stick unsalted butter; 1 slice smoked bacon, crisped and chopped; 1 teaspoon reserved bacon drippings, 1 tablespoon bourbon; 1 tablespoon pure maple syrup; 1 teaspoon brown sugar; kosher salt

Herb-Lemon Zest Butter (Fish, meats, chicken, eggs, potatoes, pasta, rice): 1 stick unsalted butter, softened, ¼ cup mixed and chopped fresh herbs; 1 teaspoon lemon zest; kosher salt optional

Anchovy Butter (steaks, pastas): ½ c. softened unsalted butter; 4 minced garlic cloves; 8 anchovies in oil, drained and minced; ½ teaspoon hot paprika; ½ teaspoon lemon juice; kosher salt

HALLOWEEN SMILES

Contributed by Viv Currell

Slice one red apple, (do not peel). Spread each slice with peanut butter and place mini marshmallows in a row to resemble teeth. Top with another slice of apple. Should resemble smiles with red lips and white teeth.

COFFEE RUB FOR BEEF, PORK OR CHICKEN

Contributed by Viv Currell

2 Tbsp. fine ground instant coffee
2 Tbsp. brown sugar
1 tsp. garlic powder
½ tsp. smoked paprika
5 Tbsp. evoo

Mix together all ingredients; spread on meat/poultry and marinate 30 minutes. Cook or grill as you desire.

PINEAPPLE STUFFING

Contributed by Terry and Sherry Sanderson

1 large can crushed pineapple (drained)
1 stick softened butter or margarine
2 eggs
¾ c. (or less) sugar
7 slices or more white bread, cubed

Cream together butter, eggs and sugar; fold in drained pineapple and bread cubes. Bake in a greased casserole dish for 30 minutes at 325-350°.

BACON JAM

Contributed by Viv Currell

½ lb. bacon, sliced into small pieces
1 medium red onion, diced into small bits
1 clove garlic, minced
½ tsp. chili powder
¼ tsp. pepper
¼ tp. Salt
⅛ tsp. cayenne
3 Tbsp. maple syrup
¼ c. apple cider vinegar
¼ c. dark brown sugar

Fry bacon till crips; remove bacon to drain on paper towel. Add onion to bacon fat in pan and cook over low heat, caramelizing for 15 minutes till dark and soft. Add garlic and spices and cook 30 seconds. Add bacon back into pan along with syrup, vinegar and brown sugar. Cook over medium heat 6-8 minutes till thickened. Keep in airtight container in fridge up to 2 weeks. Good on burgers, cream cheese, baked potatoes, omelet topping, etc.

PASTA

BACON-LEEK PASTA

Contributed by Michon Chambon

1 large leek, chopped
½ lb. bacon
¼ c. evoo
1 c. white wine
1 pint whipping cream
salt and pepper
1 lb. bowtie pasta

Boil pasta according to package directions; rinse and drain. Cook bacon in extra virgin olive oil and add chopped leek. Combine pasta with bacon and leek. Add wine, whipping cream, salt and pepper.

Yield: 4

RICOTTA GNOCCHI

Contributed by Rebecca Baruffi

1 lb. ricotta
½ c. parmesan cheese grated
2 eggs
1 tsp. salt
1 tsp. pepper
1 c. flour

Combine cheese, eggs, salt and pepper. Add flour and knead until combined. Add extra flour if dough is sticky. Knead to form a soft dough. Divide into 3 or 4 pieces, and roll into ½-inch thick ropes on a floured surface. Cut each rope into 1-inch pieces. Roll lightly with fingertip on a fork to create a dimple in the gnocchi to hold the sauce.

Bring a large pot of salted water to boil. Boil Gnocchi until they float to the surface, about 1 to 2 minutes.

Drain pasta and serve with your favorite sauce.

Yield: 4 servings *Prep Time: >30 minutes* *Total Time: 75 minutes*

EASY MAC & CHEESE

Contributed by Kendall White

1 (15-oz.) jar Alfredo sauce
2 c. shredded sharp cheddar cheese
1 c. shredded Swiss cheese
¼ c. crumbled blue cheese
1 c. whole milk
¼ tsp. salt
¼ tsp. pepper
8 oz. (3 c.) cooked macaroni
Bacon bits

For sauce: in a large pot on low heat, combine half of cheddar cheese with all of the Swiss and blue cheeses, milk, salt and pepper. When cheese is melted and blended, stir in cooked macaroni. Pour all into a 2-quart baking dish; top with remaining cheddar and sprinkle with bacon bits. Bake at 400° for 20-25 minutes.

FETTUCCINE ALFREDO

Contributed by Freda Smith

1 lb. frozen fettuccine noodles
1 stick of butter
3 c. heavy cream
3 c. grated pecorino Romano
¼ c. chopped parsley plus extra for topping

Boil fettuccine; drain. In large pot, melt butter over medium heat; add cream and stir constantly. Add pecorino Romano. Keep stirring while adding ¼ c. chopped parsley, simmering for 5 minutes. Turn off heat and add pasta to sauce. Place in large serving bowl, sprinkle with more pecorino Romano and more parsley. Mangia!

PENNE WITH VODKA SAUCE

Contributed by Joanne Valencia

12 oz. penne pasta
1 28 oz. canned tomatoes
1 tbs. unsalted butter
2 shallots, minced
1 garlic clove
¼ tsp. red pepper flakes
½ c. vodka
⅔ c. heavy cream
½ c. freshly ground parmesan
Handful of fresh basil

Bring a large pot of water to boil. Add the penne and cook as directed. Reserve ½ cup of the pasta water then drain pasta. Empty the tomatoes into a bowl and crush them with your hands.

Meanwhile, melt the butter in a large skillet over medium heat. Add the shallots and cook, stirring occasionally, until slightly soft, about 3 minutes. Add the garlic and red pepper flakes and cook, stirring, 30 seconds. Remove from heat and stir in the vodka, tomatoes, and salt to taste. Return the skillet to medium heat and simmer, stirring often until the alcohol cooks off, about 7 minutes. Stir in heavy cream and cook until the sauce thickens slightly, about 3 minutes.

Add the pasta to the sauce and toss to combine, adding some reserved cooking water to loosen, if needed. Season with salt. Serve topped with more parmesan and basil.

Yield: 4 *Prep Time: >10 minutes* *Total Time: 35 minutes*

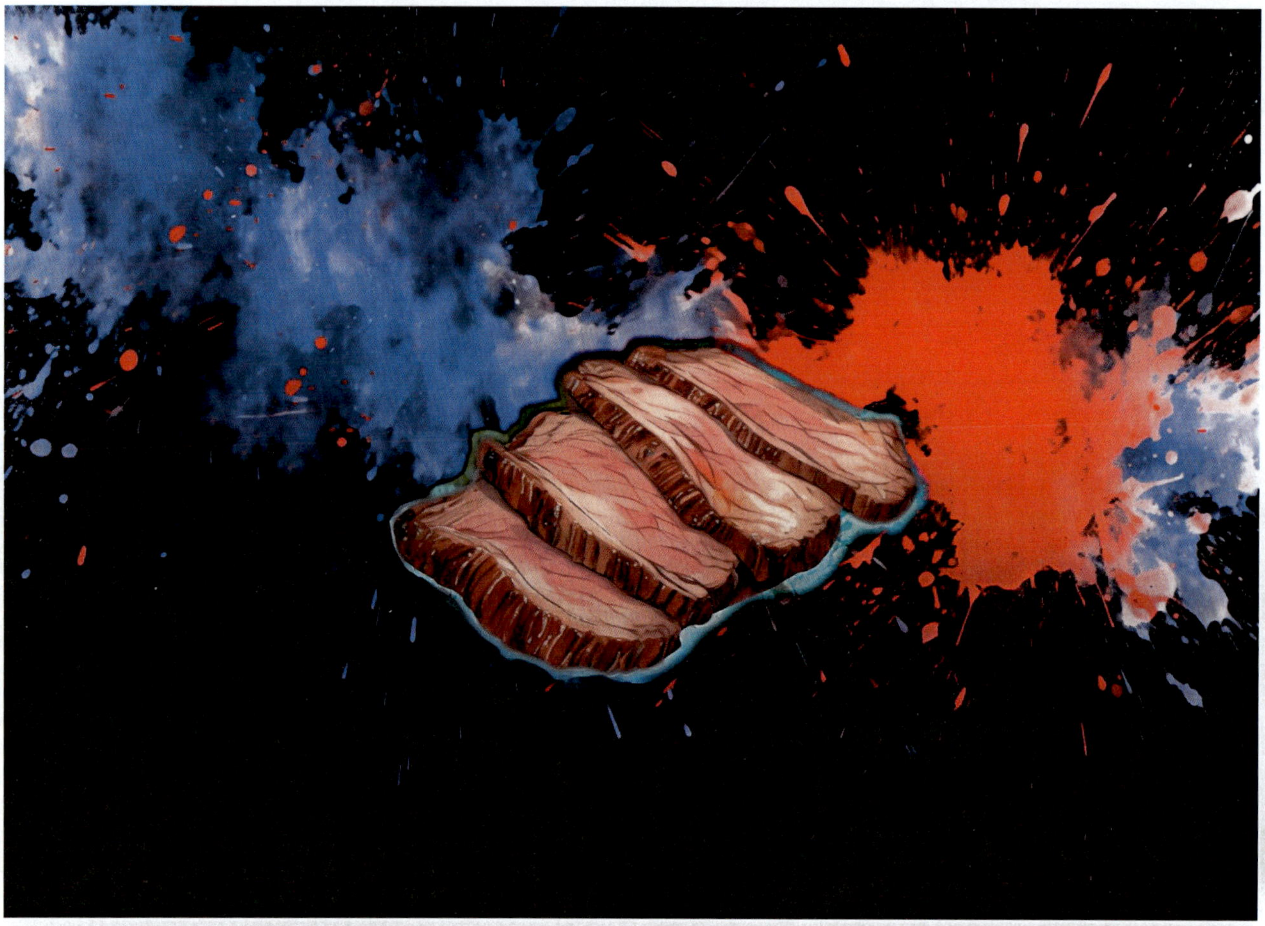

MEATS

GARLIC LOVERS POT ROAST

Contributed by Christina Eisenberg

3 lbs. boneless beef chuck roast
10 cloves garlic, peeled and cut in half
2 tsp. salt
½ tsp. ground black pepper
2 Tbsp. vegetable oil
1 c. chopped onion
1 c. peeled and chopped carrots
1 c. chopped celery
1 c. chopped celery
⅔c. water
½ to ¾ c. red wine
Rice pilaf

Using a sharp paring knife, make 20 even small slits 1 ½-inch deep all over the pot roast. Using your fingers, insert the garlic cloves as deep into the slits as possible. Season the roast on all sides with salt and pepper. In a Dutch oven, heat the oil over high heat. Carefully add the meat. Sear the meat on all sides until very well browned, 4 to 6 minutes per side. Add vegetables, water and ½ c. dry red wine. Cover and cook 3-4 hours turning roast several times to cook evenly. Add wine or water if liquid gets too low. Transfer to platter and let rest 10 minutes. Slice and serve with rice pilaf.

BEEF STROGANOFF

Contributed by Christina Eisenberg

3 Tbsp. butter
1 lb. round steak
1 (3-oz.) can mushrooms plus liquid
2 oz. water
1 ½ oz. catsup
1 c. sour cream
2 Tbsp. flour
salt and pepper
spices of choice
1 Tbsp. Sherry (optional)
3 c. cooked rice

Cut round steak diagonally in ½ inch strips about 1-inch long. Brown quickly in hot butter. Season while cooking. Add sherry, mushrooms, 1 ounce water and catsup. Bring just to a boil; add flour mixed with remaining 1 ounce of water, and sour cream. Stir until sauce just begins to thicken. Remove from heat and stir occasionally. Serve with cooked rice. Serves 4.

MARINATED FLANK STEAK

Contributed by Vivian Currell

Rub flank steak with 2 cloves crushed garlic. Squeeze juice from 1 orange and 2 limes onto steak. Sprinkle with salt and pepper. Place steak, garlic, and citrus in large baggie. Add 2 tablespoons vinegar and 2-3 tablespoons extra virgin olive oil. Seal baggie; shake to coat the steak and refrigerate 30 minutes. Grill 5 minutes on each side; cut cross grain diagonally.

ROAST BEEF BRISKET

Contributed by Ella Swinarski

4-8 lbs. brisket
Salt, pepper and Accent
Dash of paprika
2 Tbsp. garlic, minced
1 sliced onion
1 cabbage
6 medium potatoes
2 c. dry sherry wine
1 qt. water

Add brisket, seasonings, garlic and onion into covered roasting pan and bake at 400° or 30 minutes. Then add 1 quart of water and cook at 325° degrees for 4 hours. Add sherry wine. Pour broth from roasting pan into a separate pot. Add to the broth, the head of cabbage, cut into wedges, and the 6 potatoes. Cook till potatoes are done.

POT ROAST

Contributed by Christina Eisenberg for Margo Roney

1 lb. beef chuck, sirloin or choice of meat
1 onion
1 pkg. Lipton Onion Soup Mix
1 stalk celery
2 carrots
potatoes

Brown meat with onions on top of range. Add Lipton onion soup; add water. Cut up celery and carrots and cook 1 ½ hours. Then add potatoes and water as needed. Serves 4. Add more vegetables if desired.

EYE ROAST

Contributed by Rebecca Baruffi

1 3 lbs. beef eye round roast

Salt and pepper to taste

Preheat oven to 500° F.

Season roast with salt and pepper place on a roasting pan or baking dish. *Do not cover or add water.* Put roast in preheated oven and lower temperature to 475° F. Roast for 21 minutes or 7 minutes per pound. Turn off oven and let roast sit in closed oven for 2 ½ hours. *Do not open the door during this time.*

Remove when internal temperature is 145° F. Cut into slices and serve.

Yield: 8-10 *Prep Time: >5 minutes* *Total Time: 3 hours*

DILLED STROGANOFF

Contributed by Chris Eisenberg for Will Maiden

1 lb. ground beef
1 c. fresh mushrooms, sliced
1 clove minced garlic
2 Tbsp. snipped parsley
½ tsp. dried dill weed
¾ c. beef broth
½ c. dry red wine
¼ c. catsup
1 c. dairy sour cream
1 Tbsp. flour
8 oz. medium noodles, cooked and drained
Grated Parmesan cheese

In 10-inch skillet, cook ground beef, onion, mushrooms and garlic till meat is browned and vegetables are tender. Drain off excess fat. Stir in parsley, dill weed, salt, pepper, beef broth, wine and catsup. Cover and simmer 10 minutes. Blend together sour cream and flour. Stir into beef mixture, cooking and stirring until thickened and bubbly. Add cooked noodles and heat through. Sprinkle with parmesan cheese. Makes 6 servings.

PEPPER STEAK

Contributed by Christina Eisenberg

2 c. water
1 Tbsp. soy sauce
1 pkg. Knorr Ranch onion soup mix
1 ½ tsp. minced fresh or ground ginger
3 Tbsp. Mazola corn oil
2 medium chopped red or green peppers
1 clove minced garlic
1 lb. flank steak cut in thin strips
cooked rice

In a small bowl, stir water, soy sauce, soup mix and ginger until blended. In large skillet, heat 2 tablespoon corn oil over medium-high heat. Add peppers and garlic. Stir-fry 2-3 minutes or until tender-crisp. Remove from skillet. Heat remaining corn oil in skillet. Add beef, half at a time and stir-fry 2-3 minutes. Remove as browned. Return beef and pepper to skillet; stir in soup mixture. Stirring constantly, bring to boil and boil 1 minute. Serve over rice. Serves 4

BEEF BURGUNDY

Contributed by Chris Eisenberg

For beef:

¼ c. flour
2 tsp. pepper
2 lbs. round steak, cut into bite sized pieces
½ tsp. salt

In plastic bag, combine flour, pepper and salt and lightly toss with meat to coat. Brown meat in nonstick skillet.

Slow Cooker:

1 can cream of mushroom soup
1 can cream of celery soup
1 pkg. dry onion soup mix
½ c. Burgundy wine

Add all ingredients to slow cooker. Do not dilute the soups. Ad browned meat; cover and slow-cook on low for 8 hours. Recipe can also be prepared in a covered roasting pan in a 325° oven for 3 hours.

Note: Tastes even better if cook has wine while cooking.

SHIRLEY'S BEEF STROGANOFF

Contributed by Shirley Rearick

Round steak
2 Tbsp. oil
2 beef bouillon cubes
1 c. chopped onions
1 c. chopped celery
1 c. water (more if needed to prevent meat from burning)
3 bay leaves
1 pinch dried red pepper flakes
½ tsp. crushed oregano
½ tsp. crushed basil
2 cloves minced garlic
1 tsp. garlic salt
¼ tsp. pepper
1 can cream of mushroom soup
1 can cream of celery soup
1-2 c. sour cream
Noodles

Cut round steak into strips (any length/width), throwing out any gristle or fat. Brown steak in oil, stirring occasionally and adding in chopped onions and chopped celery as it browns. When browned on the outside, add the water, bouillon cubes, bay leaves, garlic salt, pepper, oregano, garlic, basil and dried red pepper flakes. Cook to boiling, stirring occasionally until steak is as tender as you like it. (up to an hour, covered). Add more water as needed to keep mixture from burning. Just before serving, add cream of mushroom and cream of celery soups. Stir well to make a gravy. When mixed well and bubbly, add 1-2 cups of sour cream. IMPORTANT: Do not continue cooking once sour cream is added. Just heat the sour cream mixture until very hot but not boiling; then turn off heat. Serve over hot cooked noodles.

CHICKEN

CHICKEN MARSALA

Contributed by Christina Eisenberg

Marsala:

½ c. butter
1 slice prosciutto, diced
2 tsp. minced shallots
2 tsp. minced garlic
2 (4 oz.) cans sliced, drained mushrooms
¼ c. Marsala wine
¼ tsp. ground black pepper
1 c. chicken stock
2 tsp. cornstarch
1 tsp. minced fresh parsley
2 tsp. heavy cream

Chicken Spices:

1 ¼ tsp. salt
1 tsp. black pepper
½ tsp. dried oregano
½ tsp. dried thyme
½ tsp. dried parsley
½ tsp. marjoram
¼ tsp. garlic powder

Use about four medium sized chicken breasts or chicken cutlets. Pound to flatten; set aside. Over low heat in medium saucepan, melt butter. Turn heat up to medium-high, add prosciutto and sauté for 2-3 minutes. Add shallots and garlic; sauté 30 seconds. Add marsala wine. Simmer 30 seconds, then add mushrooms. Drain off liquid and add black pepper. Simmer over medium heat for 5 minutes. Dissolve cornstarch in chicken stock and add to saucepan. Simmer 5 minutes. Add parsley and cream; simmer 3-4 minutes until thickened. Remove pan from heat and cover. Set aside. Preheat barbecue grill on high heat. Combine ingredients for Chicken Spices in a small bowl, using your fingertips to crush the spices in the bowl. Brush each chicken breast generously with olive oil. Sprinkle spice blend over both sides of chicken and grill for 8 minutes per side. Give chicken a turn on each side while cooking to make grill marks. Transfer to serving plate; spoon one quarter of marsala sauce over each serving of chicken.

CRAB-STUFFED CHICKEN BREASTS

Contributed by Christina Eisenberg

6 (3 oz.) chicken breast (boneless/skinless)
3 Tbsp. mayonnaise
¼ tsp. salt
⅛ tsp. black pepper
¼ tsp. seafood seasoning
1 lb. crabmeat
3 slices Swiss cheese, cut in half
3 slices boiled ham, cut in half
⅔ c. flour
1 egg mixed with a little water
½ bread crumbs mixed with ½ tsp. each of pepper and paprika
1 Tbsp. parsley
4 Tbsp. oil

Preheat oven to 400°. Pound chicken to flatten, thin enough to roll. In a bowl, combine mayo and seasonings. Blend well. Stir in crabmeat and place about ¼ to ⅓ cup of crabmeat mixture on one end of each chicken breast. Top filling with ½ slice of cheese and ½ slice of ham. Roll up and seal ends. Dip each roll in flour then egg wash and bread crumb mixture. Heat 2 Tbsp. oil in large skillet and brown chicken breasts on both sides. Transfer to oiled baking dish; sprinkle with paprika and bake 10 minutes.

Yield: 6

STUFFED CHICKEN BREAST

Unknown Contributor

2 boneless chicken breasts
2-3 c. fresh spinach
⅓ c. cream cheese (or ricotta)
Garlic powder, salt, pepper & paprika, to taste
¼ c. grated cheese-any kind

Drizzle a little olive oil in pan and cook spinach on medium heat till wilted. Season spinach with salt and pepper. Add cream cheese and mix. Cut slits into chicken breasts, season with salt pepper and garlic powder. Stuff spinach mixture into each slit. Top chicken with grated cheese; sprinkle with paprika and bake 30-40 minutes at 400°.

CHICKEN BROCCOLI CASSEROLE

Contributed by Terry and Sherry Sanderson

1 pkg. boneless chicken breast (boiled and chunked)

1 jar Cheese Wiz

1 can chicken broth

1 can cream of chicken soup

1 pkg. frozen chopped broccoli

½ chopped onion

1 bag Boil in Bag rice (or equivalent Minute Rice)

Mix all ingredients together in a bowl; spoon into greased pan and cook for 1 hour at 375° F.

CHICKEN FRANCHESE

Contributed by Freda Smith

4 skinless chicken breasts, cut in half lengthwise

Flour

Salt & Pepper

Vegetable Oil

½ lb. butter

½ bottle white wine

2 lemons

2 eggs

Pound each breast until thin. Dredge in flour, egg batter and flour again. Take large skillet and heat vegetable oil (1-inch deep), to 350° degrees. Fry chicken pieces 2 minutes on each side to lightly brown. Add lemon slices on top. Remove from pan and pat to dry. In separate skillet, add butter and wine. Simmer with lemon slices. Add chicken into mixture and cook 6 minutes till chicken is fully cooked. Serve with spaghetti.

ALIEN

Contributed by Viv Currell

1 pkg. Stove Top Stuffing
1 can cream of mushroom soup
1 whole fryer, cut up
¼ tsp. paprika
¼ tsp. each – salt and pepper
Pam or other cooking spray

Prepare Stove Top Stuffing according to package directions. Spray bottom and sides of casserole dish with cooking spray. Spread stuffing in bottom of casserole dish. Sprinkle chicken pieces with salt and pepper; place on top of stuffing. Spread mushroom soup over chicken pieces and sprinkle with paprika. Bake at 350° till chicken is done—about 45 minutes.

FISH & SEAFOOD

SHRIMP NOODLE BAKE

Contributed by Sandy Coughlin

2 c. raw medium noodles
1 can cream of shrimp soup
¾ c. mayonnaise
½ c. milk
6-8 par-cooked shrimp, chopped

Mix all ingredients together. Put in casserole dish and top with grated cheddar cheese (8 oz.) Place in preheated 350° oven for 35 minutes. 10 minutes before done, top with onion rings and bake for final 10 minutes. Serves 4.

COCONUT SHRIMP

Contributed by Christina Eisenberg

1 small red onion
4 scallions, sliced
2 cloves garlic, chopped
1 small red pepper, chopped
1 tsp. olive oil
½ tsp. dry oregano
¼ tsp. cumin
1 small can diced tomatoes
1 chicken bouillon cube
1 lb. shrimp, scallops or firm fish, cubed
1 can coconut milk
1 tsp. cornstarch (optional-for thick sauce)
1 Tbsp. fresh cilantro, chopped
Salt and pepper to taste
Hot sauce to taste

Sauté first 4 ingredients in oil until soft and transparent. Add oregano and cumin. Mix and sauté for ½ minute. Then add tomatoes and chicken bouillon cube. Cook until softly boiling (simmer). Then add seafood. Cook until shrimp are pink (if shrimp is your choice). Add coconut milk and cornstarch. Heat to simmer (DO NOT BOIL). Add cilantro. Adjust remaining seasoning ingredients to taste. Serve over fluffy white rice. (Native dish from Ecuador)

PEPPER SEARED YELLOW FIN TUNA

Contributed by Denise Bonadeo

4 tuna steaks
4 oz. goat cheese
1 oz. sundried tomatoes
8 shitake mushrooms
8 large basil leaves
2 garlic cloves
3 tbs. olive oil
1 ½ oz. olive oil

In a hot saucepan heat ½ ounce olive oil. Place mushroom caps top side down in the pan. Sear until slightly crisp turn and sear other side briefly. Remove from heat and set aside.

Preheat oven to 400° F.

Assemble tuna steaks. Open butterflied tuna steaks and place ingredients on bottom flap in this order: 1 ounce goat cheese, ¼ ounce sundried tomato strips, 2 large shitake mushrooms, and basil leaves. Brush the inside flap with garlic and oil. Close tuna steaks and salt and pepper.

Heat a heavy gauge, cast iron oven-proof sauté pan large enough to hold four tuna steaks (10-12-inches). When hot, add remainder ounce of oil, and heat till shimmering. Carefully place the 4 tuna steaks in heated oil. Sear till brown and carefully turn over with a large spatula.

Place in heated oven and bake for 5-10 minutes. Place on plates and serve warm.

Yield: 6 *Prep Time: >10 minutes* *Total Time: 45 minutes*

DESSERTS

GRANDMA BONFIETTI'S APPLE CAKE

Contributed by Joanie Thomas

1 ½ c. sugar
1 stick unsalted butter
½ tsp. salt
2 eggs
1 tsp. vanilla
1 tsp. Cinnamon
1 ½ c. flour
1 tsp. baking soda
1 tsp. baking powder
4 c. apples diced
1 c. walnuts chopped

Cream together sugar, salt and softened butter. Add 2 eggs and vanilla.

In a large bowl combine the flour, cinnamon, baking soda and baking powder. Slowly add the wet ingredients to dry using a wooden spoon.

Fold in apples and walnuts (if desired).

Pour into a prepared 13 x 9-inch baking pan.

Bake at 350° for 35-40 minutes.

Yield: 8-10 *Prep Time: >15 minutes* *Total Times: 55 minutes*

BRAZIL TORTE

Contributed by Ella Swinarski

6 eggs, separated
1 c. sugar
1 tsp. vanilla
2 lbs. Brazil nuts, ground
pinch of salt
Sweetened whipped cream

Beat egg whites until dry. In separate bowl, beat egg yolks until frothy. Add sugar a little at a time to the yolks. Keep beating; then take a good spoonful of beaten egg whites and blend with the yolks. Add nuts, folding into egg yolks carefully. Then carefully fold in remaining egg whites. Bake in two 8-inch cake pans at 300° for 35-40 minutes. Let cakes cool. Put together with sweetened whipped cream. Even better baked the day before.

GRANDMA'S CHERRY SQUARES

Contributed by: Barb Hoertz

¾ c. sugar
2 sticks of oleo
4 ½ c. flour
2 tsp. baking soda
1 tsp. vanilla
1 c. sour cream
3 egg yolks
2 cans of cherry filling

Preheat oven to 350° F. Prepare a cookie sheet that has sides on it with cooking spray.

Cream together the sugar and oleo. Add the remaining ingredients except the cherry pie filling. After dough is formed reserve some to form the top lattice. Take the remaining dough and press it into the prepared cookie sheet. Pour cherry filling on top of the pressed dough. Put the lattice on top overlapping every other strip. Use egg whites to brush on top of the lattice and bake till golden brown. Cut into squares and serve.

Yield: 12 *Prep Time: >40 minutes* *Total Time: 50 minutes*

GRAHAM CRACKER SQUARES

Contributed by Terry and Sherry Sanderson

1 can sweetened condensed milk
2 c. crushed graham cracker crumbs
1 small package chocolate chips

Mix all ingredients together; spread in greased pan. Bake at 350° for about 20 minutes. Cool slightly, cut into squares and roll in powdered sugar. YUM!

FROZEN MARGARITA PIE

Contributed by Viv Currell

Crust:

½ c. plus 2 Tbsp. butter

5 oz. salted pretzels crushed fine (1-1/4 c.)

½ c. sugar

Filling:

1 can (14-oz.) sweetened condensed milk

1 ½ Tbsp. fresh lime juice

1 ½ Tbsp. tequila

1 Tbsp. plus 1 tsp. triple sec

1 drop green food coloring

2 ½ c. whipping cream

Generously grease 9-inch pie pan. Melt butter in medium saucepan over low heat. Remove from heat; stir in pretzel crumbs and sugar till well blended. Press mixture firmly over bottom and sides of pie pan forming a rim above the pan edge. Place in freezer to firm while making filling. For filling, mix condensed milk, lime juice, tequila, triple sec and food coloring in bowl. In separate bowl, beat whipping cream till stiff peaks form. Gradually add condensed milk mixture into whipping cream a little at a time and continue beating till all is mixed and peaks still form. Pour filling into frozen pie crust. Freeze overnight. Before serving, garnish with crushed pretzels sprinkled lightly on top of pie.

BANANA SPLIT DESSERT

Contributed by Joanne Valencia

Crust

12 full graham crackers

9 Tbsp.. melted butter

1/3 c. granulated sugar

Filling

8 oz. cream cheese softened.

3 c. powdered sugar

4 medium bananas sliced

20 oz. crushed pineapple drained

16 oz. Cool Whip

1 jar maraschino cherries

½ c. chopped walnuts

Hot fudge sauce

Prepare a 9 x 13-inch pan with nonstick spray. In a medium bowl combine graham cracker crumbs, butter, and sugar. Mix until crumbs are all coated, and the mixture sticks together. Press into the baking pan. Set it aside.

Meanwhile, in another bowl stir together the cream cheese, ¼ cup of butter and powdered sugar. Mix until creamy. Scoop the cream cheese mixture onto the graham cracker crust. Press on the bananas evenly, then the pineapple, and top with the Cool Whip.

Cool until serving. Enjoy!

Yield: 4 *Prep Time: >10 minutes* *Total Time: 35 minutes*

BETTER THAN SEX DESSERT

Contributed by Chris Eisenberg

1 box chocolate cake mix
1 large box chocolate pudding
4-5 regular Heath candy bars, crushed
1 pkg. Cool Whip

Prepare cake according to box directions. Cool. Make pudding according to directions. In bowl, place one layer of cake and top with half of the pudding. Sprinkle a third of the crushed candy. Add a layer of Cool Whip. Repeat with remaining cake, pudding, candy and Cool Whip. Sprinkle any remaining candy on top. May also use yellow cake with vanilla pudding, or yellow cake with chocolate pudding, or chocolate cake with vanilla pudding.

BROWNIES

Contributed by Ella Swinarski

4 squares unsweetened chocolate
1 ½ stick butter
2 c. sugar
4 eggs
1 c. flour
½ tsp. salt
1 tsp. vanilla
½ lb. chopped pecans

Mix all ingredients and pour into a 10-inch square baking pan. Bake at 325° for 35-40 minutes.

BROWNIE

Contributed by Rebecca Baruffi

4 squares unsweetened chocolate
1 ½ sticks of butter
4 eggs
1 cup flour
½ tsp. salt
1 tsp. vanilla
½ lb. chopped pecans

Mix all ingredients and pour into a prepared 10-inch square pan.

Bake at 325° F for 35-40 minutes.

Yield: 12 *Prep Time: >15 minutes* *Total Time: 55 minutes*

FROSTED STRAWBERRY SQUARES

Contributed by: Barb Hoertz

1 c. flour
¼ c. brown sugar
½ c. chopped nuts
½ c. melted butter
1 tbs. lemon juice
¾ c. sugar
3 egg whites
2 cans of cherry filling
1 10 oz. pkg. frozen strawberries
1 frozen Cool Whip

Preheat oven to 350° F. Prepare a 13 x 9-inch pan.

In a bowl stir flour, brown sugar, nuts, and butter. Spread in prepared pan and bake for 20 minutes, stir occasionally. Remove from the pan and let cool. Sprinkle ⅔of the cooled mixture back into the same pan. Reserve remaining ingredients.

In a large bowl, combine egg whites, berries and lemon juice. Beat at low speed until mixture begins to thicken (about 2 minutes.) Beat on high until very stiff (10-12 minutes). Fold in the Cool Whip. Spread over the mixture in the pan. Top with remaining crumbs. Freeze 6 hours or overnight.

Yield: 12 *Prep Time: >10 minutes* *Total Time: 30 minutes*

FUNFETTI COOKIES

Contributed by Michelle Divine and Rebecca Baruffi

1 box funfetti cake mix
¾ c. white chocolate chips
1/3 c. oil
2 eggs, beaten

Combine all ingredients and mix until well blended. Use a scoop or roll into balls. Place on a baking sheet apart and slightly flattened.

Bake at 350° F for 10-12 minutes.

Yield: 3 dozen *Prep Time: >10 minutes* *Total Time: 35 minutes*

CRUMB CAKE FOR A CROWD

Contributed by Rebecca Baruffi

For the cake

1 box of yellow cake mix
3 eggs
1 ⅓ c. water
1 tbs powdered sugar (optional)

For the crumb topping

3 tsp. cinnamon
4 c. flour
1 lb. softened butter
1 pinch of salt
1 ⅓ c. sugar
3 tsp. vanilla

Combine cake ingredients and beat for 4 minutes. This will create a very satiny, thick batter.

Spread in a 12 x 17-inch prepared jelly roll pan.

Bake at 350° F for 10 minutes.

Combine crumb topping ingredients until it has the consistency of wet sand. Crumble over partially cooked cake. Return cake to oven for an additional 10-15 minutes (to test for doneness: an inserted toothpick should come out clean). Cool and sprinkle with powdered sugar (optional).

Yield: 20 *Prep Time: >10 minutes* *Total Time: 35 minutes*

MONSTER COOKIES

Contributed by Denise Gauthier

⅔ c. butter
1 ¼ c. brown sugar
¾ c. white sugar
3 beaten eggs
1 ½ c. peanut butter
6 c. oatmeal
2 tsp. baking soda
1 c. raisins
12 oz. semi-sweet chocolate chips

Steam and cool raisins to plump them. Melt butter. Blend until smooth, the sugars, eggs and peanut butter. Mi all remaining ingredients into very large bowl. Drop onto greased cookie sheets. Bake at 350° for 10-12 minutes, depending on size of cookies. Yields approximately 7 dozen

NO BAKE PINEAPPLE CHEESECAKE

Contributed by Livia Coates

2 pks. (8-oz.) cream cheese, softened
½ c. sugar
1 can (20-oz.) crushed pineapple, drained
2 pkgs. lady fingers
1 pint fresh strawberries

In a large mixing bowl, beat the cream cheese and sugar until smooth. Stir in pineapple. Fold in whipped topping. Place lady fingers around the sides and bottom of a greased 9-inch spring form pan. Pour cream cheese mixture into pan. Cover and refrigerate at least 8 hours or overnight. Remove sides of pan and top cheesecake with sliced or halved strawberries.

PEANUT BUTTER PIE

Contributed by Denise Gauthier

Crust

1 ¼ c. chocolate cookie crumbs (or a store-bought Oreo pie crust)

¼ c. sugar

¼ c. melted butter

Filling:

8 oz. softened cream cheese

1 c. creamy peanut butter

1 c. sugar

1 Tbsp. softened butter

1 tsp. vanilla extract

1 c. heavy cream, whipped

Grated chocolate or cookie crumbs

Combine crust ingredients and press into a 9-inch pie plate; bake at 375° for 10 minutes and let cool. In a mixing bowl, beat cream cheese, peanut butter, sugar, butter and vanilla until smooth. Fold in whipped cream. Gently spoon into pie crust. Garnish with chocolate or cookie crumbs if desired. Refrigerate; serve chilled.

STRAWBERRY PIE

Contributed by (Unknown)

1 – 8-inch graham cracker crust

1 qt. fresh strawberries

1 small vanilla pudding (cook and serve)

1 small strawberry Jell-o

1 ½ c. water

Whipped Cream or Cool Whip

Slice strawberries and pour into crust. Mix pudding and Jell-o in pan with water, stirring constantly till boiling and thickened. Pour over strawberries; cool and top with whipped cream (refrigerate).

CHEESECAKE

Contributed by Ella Swinarski

Crust:

2 c. graham cracker crumbs
½ c. sugar
⅓ c. softened butter

Filling:

1 c. sugar
½ tsp. salt
1 c. eggs (4-5 large eggs)
3 Tbsp. canned milk
⅓ c. lemon juice
¼ c. flour
4 pkgs. Cream cheese (8-oz. each)
½ tsp. vanilla

Grease spring form pan. Mix together crust ingredients and press mixture into bottom and ¾ way up on sides of pan. Set aside. Combine sugar, flour and salt. Beat cream cheese. Alternately add to cream cheese, eggs and flour, mixing to combine with each alternating addition. Add in vanilla, lemon and canned milk. Pour into graham cracker crust and bake 40 minutes at 375°. Cool. Thicken fruit juice and decorate top of cheesecake, adding fresh or canned fruit.

PECAN BROWNIES

Contributed by Ella Swinarski

4 squares unsweetened chocolate
1 ½ sticks of butter
4 eggs
1 c. flour
½ tsp. salt
1 tsp. vanilla
½ lb. chopped pecans

Mix all ingredients and pour into a prepared 10-inch square pan.

Bake at 325° F for 35-40 minutes.

Yield: 12 *Prep Time: >15 minutes* *Total Time: 55 minutes*

HOMEMADE CINNAMON ROLL CAKE

Contributed by: Joanne Valencia

For the Cake

3 c. all-purpose flour
1 c. white granulated sugar
4 tsp. baking powder
2 large eggs
2 tsp. vanilla extract
1 ½ c. milk
½ c. salted butter 1 stick (melted)

For the Cinnamon Filling

¾ c. salted butter softened
1 c. light brown sugar
2 tbs. flour
1 tbs. ground cinnamon

For the Glaze

2 c. powdered sugar
5 Tbsp. milk
1 tsp. vanilla extract

Preheat oven to 350° F. Spray a 9 x 13-inch baking dish with nonstick cooking spray. Set it aside.

In a large bowl combine the ingredients for cake (except melted butter). Just stir until combined. Do not overmix. While your mixer is running slowly add the melted butter until combined. Spread the batter in the prepared pan.

In a small bowl combine all your filling ingredients. Stir until combined well. Drop by spoonful onto the cake batter. Take a butter knife and make inserted comes out clean.

For the glaze, in a small bowl whisk together powdered sugar, milk and vanilla. Pour the glaze over the warm cake.

Yield: 12 *Prep Time: >40 minutes* *Total Time: 50 minutes*

EASY CHEESECAKE

Contributed by Barb Hoetz

Filling

6-8 oz. cream cheese

1 tsp. vanilla

2 ½ c. sugar

6 eggs

Crust

½ lb. graham cracker crumbs

4 Tbsp. sugar

3 tsp. cinnamon

12 Tbsp. melted butter

Topping

4 c. sour cream

10 Tbsp. sugar

1 tsp. vanilla

Preheat oven to 350° F. Prepare a cheesecake pan.

In a bowl cream together cream cheese, vanilla and 2 ½ cups of sugar. Add 6 eggs to creamed mixture.

For the crust, mix together the graham crackers, sugar, cinnamon and butter. Bake till the top is firm.

Prepare the topping by mixing the sour cream, 10 tablespoons of sugar and vanilla. Spread over the cooked cheesecake, and bake at 500° F for 5 minutes.

Yield: 12 *Prep Time: >10 minutes* *Total Time: 30 minutes*

BREADS, MUFFINS & ROLLS

HIGBIE'S IRISH BREAD DELIGHT

Contributed by Chris Eisenberg for Lollie Higbie

3 c. flour
1 Tbsp. baking powder
⅓ c. white sugar
1 tsp. salt
1 tsp. baking soda
1 egg, slightly beaten
2 c. buttermilk
¼ c. melted butter
1 c. raisins
½ c. golden raisins
½ c. dried currants

Mix well by hand and pour into greased 9 x 5-inch loaf pan. Bake at 325° for 65-75 minutes. Cool on rack; then wrap in foil.

IRISH SODA BREAD

Contributed by Chris Eisenberg for Lollie Higbie

2 c. unsifted flour
½ c. sugar
2 tsp. baking powder
1 tsp. baking soda
1 tsp. salt
1 c. buttermilk
2 c. raisins (1 c. white/1 c. dark)

Preheat oven to 375°. Lightly grease a small cookie sheet. In a large bowl, sift flour, sugar, baking powder and soda, salt and raisins. Add buttermilk. Mix with a fork only till dry ingredients are moistened. Turn out onto lightly floured board and knead gently until smooth, about 1 minute. Shape into a ball and place on prepared cookie sheet. Flatten into 7-inch circle. Press a large floured knife into the center of loaf almost through to bottom. Repeat 2 right angles to divide loaf into quarters. Bake 30-40 minutes or until top is golden and loaf sounds hollow when tapped. Remove to wire rack to cool or it will get mushy on bottom. Best of health. Excellent with a spot of Irish tea.

ZUCCHINI BREAD

Contributed by Chris Eisenberg for Will Maiden

2 c. grated zucchini
3 eggs
2 c. sugar
1 c. oil
1 tsp. vanilla
3 c. flour
3 tsp. cinnamon
¼ tsp. baking soda
½ c. chopped nuts

Wash and dry zucchini; do not peel or cook. In large bowl, beat eggs. Add sugar, oil and vanilla; mix well. Add flour and remaining dry ingredients, alternating with the squash. Mix well and stir in nuts. Bake at 350° for 60-75 minutes. Makes one loaf.

CRANBERRY NUT BREAD

Contributed by Viv Currell

1 ½ c. whole raw cranberries
½ c. sugar
2 c. sifted flour
¾ c. sugar
1 ¼ tsp. baking powder
½ tsp. salt
½ tsp. baking soda
1 orange
¼ c. butter or Crisco
2 eggs
½ c. chopped walnuts

In small saucepan, combine cranberries and ½ c. sugar. Cook over medium heat till sauce bubbles (8-10 minutes); set aside to drain. Sift together flour, remaining ¾ cup sugar, baking powder, salt and soda. Grate peel of orange. Squeeze juice into measuring cup. Add grated peel and enough water to juice to make ¾ cup liquid. Add mixture to dry ingredients with butter and eggs. Mix well. Fold in nuts and drained cranberries. Pour into greased loaf pan. Bake at 350° for 45-50 minutes. Remove from oven and while still warm, wrap in saran wrap or waxed paper and then in foil. Freeze overnight and slice thin while still frozen. Serve with whipped cream cheese.

BANANA NUT BREAD

Contributed by Denise Gauthier

1 ½ c. flour
1 c. sugar
1 stick butter
1 c. walnuts, chopped
2 mashed bananas
2 eggs
pinch of salt
½ tsp. baking soda

Cream together butter and sugar. Add remaining ingredients and mix with a mixer until well blended. Pour into a greased loaf pan and bake one hour at 350°. Cover with foil around edges if they are getting too done. May work best cooked on lower oven rack.

REFRIGERATOR POTATO ROLLS

Contributed by Chris Eisenberg for Will Maiden

2 dried yeast cakes
1 c. butter or shortening
1 ½ c. sugar
3 tsp. salt
2 c. mashed potatoes
2 c. water (from mashed potatoes)
3 eggs
5-6 c. bread flour
Water

Dissolve yeast in lukewarm water. Check to make sure yeast is active. Put potato water into large bowl and add shortening, sugar, salt and mashed potatoes. When cold, add yeast and slightly beaten eggs. Sift in enough flour to make a stiff dough. Turn out on flour and knead well. Put in greased bowl. Grease top of dough; cover tightly. Place in refrigerator overnight. Form into desired shape, let rise and bake at 375°. Makes about 30 rolls.

BARBECUE

BARBECUE GRILL RANCH RIBS

Contributed by Viv Currell

6 lbs. spare ribs
⅔ c. Karo all-purpose syrup
2 Tbsp. corn starch
⅓ c. soy sauce
¼ tsp. salt; dash of pepper
1 clove crushed garlic
½ tsp. ginger
¼ c. vinegar

Combine all ingredients in a small saucepan; bring to a boil; pour over spareribs. Cover and refrigerate four hours, turning occasionally. Grill slowly, basting often.

Note: Spare ribs are done when they reach an internal temperature between 200-202° F.

EASY BARBECUE SAUCE FOR PORK AND CHICKEN

Contributed by Viv Currell

⅔ c. ketchup
½ c. cider vinegar
¼ c. brown sugar
2 tsp. smoked paprika
1 tsp. ground cumin
1 tsp. kosher salt
1 tsp. freshly cracked black pepper
Dash of hot sauce (optional)

Combine all ingredients in a small saucepan; bring to a simmer over medium heat and cook for 5 minutes.

Note: To prevent burning if cooking on the grill, thin out sauce with a little bit of water before spreading on chicken or pork.

Yield: 1 ½ cups

STEP ON A CRACK; BREAK YOUR MOTHER'S BABY BACK RIBS

Contributed by Viv Currell

For the ribs:

5 lbs. baby back pork ribs

2 c. dark beer

2 c. orange juice

2 c. apple cider vinegar

For dry rub:

1 Tbsp. each of fresh cracked black pepper, ground coriander and ground thyme

2 Tbsp. each of brown sugar, celery salt and chili powder

3 Tbsp. each of paprika (I like smoked paprika), and garlic powder

For Sauce:

2 Tbsp. butter

1 finely chopped medium sized onion

4 cloves finely chopped garlic

1 tsp. dry mustard

3 Tbsp. brown sugar

1 tsp. instant coffee

½ tsp. chipotle chili sauce or hot pepper sauce

2 c. tomato puree

½ c. apple cider vinegar

Preheat oven to 450° F. While oven heats, place ribs in a pot on the stove over high heat. Cover with beer, orange juice, vinegar and just enough water to submerge ribs in liquid (approximately 2 cups). When liquid beings to boil, cover pot and remove from heat; set aside for 20 minutes. To prepare sauce, melt butter In a saucepan over medium heat, add onions and garlic and sauté till tender. Add mustard, brown sugar and coffee to saucepan and cook for one minute, stirring constantly. Stir in chili sauce, tomato puree and vinegar. Bring to a boil, then reduce heat and simmer for at least 30 minutes. Season to taste with salt, pepper and hot sauce. In a bowl, combine dry rub ingredients. Remove ribs from pot; drain and lightly pat dry, leaving some moisture to absorb dry rub. Rub well with dry rub mixture. Place ribs, fat side up in a foil lined pan and bake 30 minutes. Halfway thru baking, turn ribs over and reduce ribs to 350° F. Degrease pan and cover ribs with sauce. Cover tightly with foil and continue baking until tender, about 45 minutes. A long process for sure, but worth it.

Note: Baby back ribs are done when they reach an internal temperature between 198-200° F, for fall-off-the-bone doneness the internal temperature should be over 200° F.

EASY BARBEQUE SAUCE FOR BEEF RIBS

Contributed by Viv Currell

1 ½ c. brown sugar
1 ½ c. ketchup
½ c. apple cider vinegar
¼ c. water
1 Tbsp. Worcestershire sauce
1 ½ Tbsp. chili powder
2 tsp. paprika (smoked is preferred)
2 tsp. salt
1 Tbsp. dry minced garlic
1 Tbsp. dry minced onion
1-1/2 tsp. black pepper
½ tsp. cayenne

Combine all ingredients in a saucepan over medium heat. Bring to a boil and quickly drop the heat to low. Simmer for 5 minutes.

HONEY BBQ SAUCE

(Adapted from Southern Living Magazine)

Contributed by Viv Currell

1 c. ketchup
⅓ c. honey
¼ c. butter (unsalted if preferred)
2 Tbsp. fresh lemon juice (1 lemon)
1 Tbsp. Worcestershire sauce
2 lemon slices

Place all ingredients (except lemon slices) into a small saucepan and bring to a boil over medium heat. Reduce heat to low and add lemon slices. Simmer for 30 minutes. Remove from heat; discard lemon slices and cool sauce completely.

MUSTARD BBQ SAUCE

Contributed by Viv Currell

¾ c. yellow mustard
½ c. honey
¼ c. brown sugar
½ c. apple cider vinegar
2 Tbsp. ketchup
2 tsp. Worcestershire sauce
1 tsp. garlic powder
½ tsp. salt
Pinch of cayenne
Dash of hot sauce

Mix all ingredients together in saucepan; simmer 10 minutes and then allow to cool. Pour into a jar with lid and store in fridge overnight or longer to allow flavors to mingle.

MY RECIPES

The next few pages are left intentionally blank for you to make this cookbook your own and add a few recipes from your kitchen...

WWII Recruiting Posters

Here are some of the most famous US military recruiting posters of World War II.

1. Perhaps the most famous poster ever was of "Uncle Sam" and while it was used extensively during World War II, it actually first came out in 1917.

2. But Sam showed up in World War II-specific recruiting efforts as well, like this one below from 1944. And the original poster can still often be seen at modern recruitment offices.

3. In the wake of Japan's surprise attack on Pearl Harbor on Dec. 7, 1941, many answered the call to "Avenge Pearl Harbor."

4. While the era's posters were not very politically correct, they were effective. It's worth noting however, that many soldiers were drafted.

5. This poster recruited men to join the "Flying Leathernecks."

6. While this one pushed for Navy enlistments. The war in the Pacific during World War II was the largest naval conflict in history, according to CombinedFleet.com.

7. This popular poster of a U.S. Marine "ready" from 1942 was so iconic, an updated version of a Marine with the tagline of "still ready" was made in the Post-9/11 era.

8. And for those on the home front, the "Rosie the Riveter" poster became well-known for motivating women to take over factory jobs men had left behind to fight in the war.

Veterans' Organizations

AMVETS

“

On July 23, 1947, President Harry S. Truman signed Public Law 216, making AMVETS the first World War II organization to be chartered by Congress.

Today, membership in AMVETS is open to anyone who is currently serving, or who has honorably served, in the U.S. Armed Forces from World War II to the present, to include the National Guard and Reserves.

As [AMVETS] moves further into the 21st century, it does so with the conviction that its focus on preserving freedom, supporting America's defenders and serving her communities remains a clear blueprint for continued service to God and country.

”

Veterans Helping Veterans (since 1948)

- Representing the interests of 20 million warriors and their families.
- A 501(c)3 veterans charity, we are the nation's most inclusive veterans service organization, open to all who honorably served.

American Legion

The American Legion was chartered by Congress in 1919 as a patriotic veterans organization. Focusing on service to veterans, service members and communities, the Legion evolved from a group of war-weary veterans of World War I into one of the most influential nonprofit groups in the United States.

Some significant dates in Legion history:

- Aug. 16, 2017: The Harry W. Colmery Veterans Educational Assistance Act of 2017 – also known as the "Forever GI Bill" because it removes time limits for veterans who wish to use it for college – is signed into law by President Donald J. Trump.
- Aug. 23, 2017: President Donald J. Trump signs the Veterans Appeals and Modernization Act of 2017 onstage at the 99th American Legion National Convention, in Reno, Nev.

- August 2017: Denise H. Rohan of Wisconsin is elected national commander, the first woman to hold the role in the Legion's history.
- June 20, 2019: The U.S. Supreme Court rules 7-2 in favor of The American Legion in its case to allow a 40-foot cross-shaped memorial in Prince George's County, Md., to stay on public property.
- July 30, 2019: President Trump signs the Let Everyone Get Involved in Opportunities for National Service – the LEGION Act – which extends recognized wartime service to cover previously unrecognized periods of U.S. military conflict, back to World War II.

VFW

"

The Veterans of Foreign Wars of the United States {VFW] traces our roots back to 1899 when veterans of the Spanish-American War (1898) and the Philippine Insurrection (1899-1902) founded local organizations to secure rights and benefits for their service."

Our Core Values:

- Always put the interests of our members first
- Treat donors as partners in our cause
- Promote patriotism
- Honor military service
- Ensure the care of veterans and their families
- Serve our communities
- Promote a positive image of the VFW

"

Ditch 501 c Organizations

Why Ditch 501c Organizations

No where in charter of veterans' organizations does it say veterans and / or military family members get to gamble and not pay taxes on winnings.

The US Senate and the US Congress saw the problem and, as usual, found the most circuitous route available to "solve the problem." Hence, the 501c of the IRS Code was born.

In my book, ***Complexity = Corruption / The US Tax System***, I quote Fareed Zakaria:

> The US tax system has degraded the entire system of American Government. Special interests pay politicians vast amounts of cash for their campaigns, and in return they get favorable exemptions or credits in the tax code.
>
> The U.S. tax system is not simply corrupt; it is corrupt in a deceptive manner. Congress is able to funnel vast sums of money to its favored funders through the tax code - without anyone realizing it. The simplest way to get the corruption out of Washington is to remove the prize that members of Congress give away: preferential tax treatment.

American Legion Post 126

The Post Commander at American Legion Post 126 in Jensen Beach, where I am an auxiliary member, gave up the post's non-profit status and converted to a for-profit corporation for tax purposes.

The remaining problem that still needs a solution - how does one track gambling losses, which are deductible from gambling winnings, in a cash only business.

Veterans passed, American Legion Post 126, Jensen Beach, Fla.

We Are The Mighty

Military Entertains | Inspires

We Are The Mighty is a veteran-led digital publisher and Emmy-award-winning media agency, owned by Recurrent Ventures. (https://www.wearethemighty.com)

Thanks to the Internet, today's military has expanded its mission to include entertainment and inspiration. The primary goal of Martin Pearl Press LLC, the free speech traditional publishing company, is to support the democratic process, by honoring veterans and military families.

Recurrent Military Team

> We know the military community, because we are the military community.
>
> Our production and marketing services are informed by authentic, apolitical, inspirational content about service that we create for all platforms: digital, film, and TV. We work with amazing brands and employ incredibly talented veterans.

 Mark Harper General Manager, USAF Veteran	 Curtez Riggs President-Events and Activation, USA Veteran	 Katherine Torres-Pummil Head of Brand Partnerships
 Tessa Robinson *We Are The Mighty*, Editor-in-Chief, Former CIA	 Marty Skovlund Editor-in-Chief, USA Veteran	 Tyler Rogoway *The War Zone*, Editor-in-Chief
 Jonathan Epstein Military Operations, USAF Legacy	 August Dannehl Head of Production, USN Veteran	 Philip Hladky Director of Sales, USMC Veteran

Epilogue

The US Supreme Court's decision to tear down the wall and let the illegal aliens in has spawned swift retaliation by 25 states and some say we will soon be fighting Civil War 2024 in the United States of America.

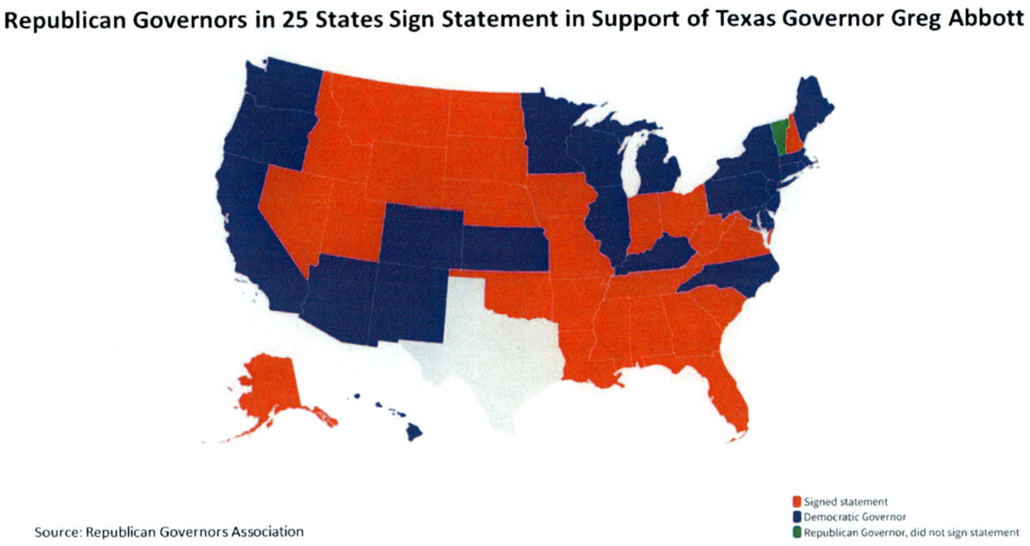

Source: https://www.usatoday.com/story/news/politics/2024/01/26/texas-border-dispute-gop-states-support-letter/72364737007/

It seems I am the only one in my crowd who doesn't have a gun and a concealed carry permit. Most of my close friends are women. They carry big purses because their guns are not small.

My male friends are all veterans. I am confident they know how to use a gun and shoot to kill, if necessary. I have heard that Florida sheriffs offered to deputize any and all local citizens who own a gun and have a concealed carry permits to shoot to kill if confronted by an illegal alien who makes threatening moves.

The situation is so crazy and fluid that I have no idea what will happen next, so this epilogue will remain incomplete for now.

Peace and Love,

Diane M. Freaney
Martin Pearl Press LLC

Events

First Annual Clay Shoot

First Annual Hobe Sound Chamber Clay Shoot at OK Corral Gun Club in Okeechobee, Florida.

The Framery Cynthia Cooper

William "Billy" Selesnick paints and writes books about the history of Martin County. Cynthia Cooper honors Billy at a Meet the Artist event at the Framery.

Resources

Books

Alien Nation: Common Sense About America's Immigration Disaster, Brimelow, Peter. Harper Perennial (1996).

Articles

"*The Increasingly Dangerous Politicization of the U.S. Military*." Barno, David and Bensahel, Nora. War on the Rocks. June 18, 2019.

"*The 8 most famous US military recruiting posters of World War II.*" By Team Mighty. *War on the Rocks.* Updated on Feb 13, 2023. https://www.wearethemighty.com/lists/world-war-2-recruiting-posters/

"Complexity Equal Corruption." Zakaria, Fareed. *Time*. Oct 31, 2011

Newsletters

Hobe Sound Beach Shop's Turtle Times

The Hobe Sound Beach Shop's Turtle Times, available free around town, tells the history of Hobe Sound, a chapter in each newsletter. Now on Volume 5, #2, one could imagine there will be no more, yet the history continues to unfold.

News You Can Use, the Hobe Sound Chamber

A weekly newsletter from the Hobe Sound which includes community events in addition to the Hobe Sound Chamber events all in one digital newsletter.

Websites

Carsons Tavern. https://www.carsonstav.com

Opened February 2007.

> "Are you looking for a local neighborhood Tavern with fantastic food and the friendliest of staffs? Carsons Tavern is your place!"

Dixie Grill Stuart Florida. https://www.dixiegrillcafe.com

Open Monday-Sunday 6:30 am-2:00 pm

Dixie Grill is nestled in the hustle and bustle of our favorite town - Stuart, Florida.

> "Stop by and try one of our breakfast or lunch options. All of our menu items are freshly prepared to order. You can dine in, eat outside on the patio, order online, or call in a pickup order! We are here to serve you."

Harry and the Natives. https://www.harryandthenatives.com

Opened on Pearl Harbor Day 1941.

> Harry MacArthur, Native Son. "I shall return."

Julie LaFrance-Lentine. https://www.stuartmagazine.com/stuart-life/people/how-julie-lafrance-lentine-made-the-taco-shack-a-must-try-stuart-restaurant/

Miami-hospitality maven Julie LaFrance-Lentine moved to Stuart and opened the thriving Taco Shack, where she serves homemade, gourmet tacos.

STUART
MAGAZINE

Stuart Life People

How Julie LaFrance-Lentine Made The Taco Shack A Must-Try Stuart Restaurant

By **Amy Lynne Hayes** - January 27, 2016

It's 3 p.m. the day after Christmas. Most of the storefronts are quiet—all except this one. The Taco Shack on Colorado Avenue is overflowing with customers—its staff hustling to take orders, clear tables and care for the customers. It's an organized hub of activity, with one person at the center pulling the strings.

Owner and founder Julie LaFrance-Lentine is passionate about the food and hospitality industry. The Miami native thrives on providing the highest quality service and food at her establishment, which celebrated its second anniversary in December.

"My goal was to create a warm, friendly, fun environment with a great vibe, great food and great service," she says. Having started in the industry at an early age, LaFrance-Lentine knows how to do just that.

Her beginnings are rooted in the hospitality scene in Miami. At the age of 15, she was working for her father, also a restaurateur. At 19, she opened the landmark Mayfair Hotel in Coconut Grove. She went on to have her hand in numerous other properties, and by the age of 26 held a position at Miami's only five-star hotel at the time, the Grand Bay Hotel. "That's when I knew I would be in the food hospitality industry," she recalls. She worked her way up, switched to the corporate side, and eventually relocated to Stuart in 1996.

When the opportunity presented itself, LaFrance-Lentine returned to her love of food and cooking. She was working in catering and events when, in 2012, Ron Hart, owner of Earthtones, invited her to cater at his outdoor bar and concert venue, Terra Fermata. So one night she bought the meat, gathered all the toppings and started serving pork tacos. Her reputation as the Taco Lady began.

"I knew I was on the verge of something, but didn't know what it was," she says.

The path revealed itself soon enough when she walked over to the under-construction office building across from Terra Fermata. A light bulb went off in her head, and the idea for a taco shop was born. In December 2013, The Taco Shack opened serving only five kinds of tacos and two soups. The restaurant now serves a full menu, with as many as 30 tacos, depending on the day.

LaFrance-Lentine's twist on gourmet tacos has now earned her the elevated name of Taco Queen. Everything served is made from scratch, in-house, from locally sourced ingredients, according to her own original recipes.

"It's a labor of love," she says smiling. The Stuart community has shown up to love her right back.

Old Dixie Cafe North. https://www.facebook.com/OldDixieCafeNorth/

> "Join us at Old Dixie Cafe North for an unbeatable Early Bird Special that's worth setting your alarm for.
>
> "Dive into 2 eggs any style, crispy hash browns, and your choice of toast with a comforting cup of coffee or soothing hot tea – all for just $6.95! 🥓🥪✨ Upgrade to ultimate satisfaction with savory sausage or crispy bacon for $7.95. Mornings made delicious, only 'til 8 AM, Mon-Fri."

Taste. https://www.tasteofhobesound.com

Taste took over the old Catfish House, which closed so the owners could travel while they were still young.

USO. https://www.uso.org

> "Since 1941, the USO has been the nation's leading organization to serve the men and women in the U.S. military, and their families, throughout their time in uniform. From the moment they join, through their assignments and deployments, and as they transition back to their communities, the USO is always by their side."

War on the Rocks. https://warontherocks.com/about/

> "War on the Rocks is a platform for analysis and debate on strategy, defense, and foreign affairs. It features articles and podcasts produced by an array of writers with deep experience in these matters: top notch scholars who study war, those who have served or worked in war zones, and more than a few who have done it all.
>
> "Among our regular contributors are people who have worked on every continent in the world. They have commanded ships, bargained with militias, led patrols, managed alliances, called for fires, and negotiated treaties. They include former diplomats, officers, NCOs, intelligence professionals, and some of the most reputable scholars in the world studying war and international politics. Our contributors include citizens of countries from around the world, members of all political parties and none. But they all converge on that one guiding principle for our project: experience.
>
> "We seek to teach people how to think about the world, rather than what to think about it. We do not impose conclusions on our authors or our readers. We reason forward to find answers, not backward from predetermined

beliefs. We believe in the centrality of fear, honor, and interest as drivers of inter-state affairs. Politics is power. À la Morgenthau, we understand power as "anything that establishes and maintains the power of man over man .... from physical violence to the most subtle psychological ties by which one mind controls another." While armed force or the threat thereof is often central to our analyses, our writers also engage deeply with ideas and social control as mechanisms for power. No topic is off limits."

Wikipedia

Book burning.

"Book burning is the deliberate destruction by fire of books or other written materials, usually carried out in a public context. The burning of books represents an element of censorship and usually proceeds from a cultural, religious, or political opposition to the materials in question.

"Book burning can be an act of contempt for the book's contents or author, intended to draw wider public attention to this opinion, or conceal the information contained in the text from being made public, such as diaries or ledgers.

The Nazi book burnings became emblematic of a harsh and oppressive regime which silenced prevailing culture."

On May 10, 1933, Nazis in Berlin burned works of Jewish authors, and the library of the Institut für Sexualwissenschaft, and other works considered "un-German".

⊘ *A member of the SA throws confiscated books into the bonfire during the public burning of "un-German" books on the Opernplatz in Berlin.*

Collaboration.

> "Collaboration is the process of two or more people, entities or organizations working together to complete a task or achieve a goal.
>
> "Teams that work collaboratively often access greater resources, recognition and rewards when facing competition for finite resources."

Jenna Bush Hager. https://en.wikipedia.org/wiki/Jenna_Bush_Hager

Jenna Bush Hager, shortly after her wedding to Henry Hager.

Jenna Bush Hager is credited with bringing the story of Alan Jackson, country star who has Charcot Marie Tooth disease, to the public's attention. Jenna Bush Hager writes children's books with her twin sister, Barbara, and her paternal grandmother, Barbara Pierce Bush.

Income tax on gambling.

> "The [United States} Internal Revenue Code contains a specific provision regulating income-tax deductions of gambling losses.
>
> "Under Section 165(d) of the Internal Revenue Code, losses from "wagering transactions" may be deducted to the extent of gains from gambling activities. Essentially, in order to qualify for a deduction of losses from wagering, the taxpayer can only deduct up to the amount of gains accrued from wagering."

United Service Organizations

The United Service Organizations Inc. (USO) is an American nonprofit-charitable corporation that provides live entertainment, such as comedians, actors and musicians, social facilities, and other programs to members of the United States Armed Forces and their families.

Founded during World War II, the USO sought to be the GI's "home away from home" and began a tradition of entertaining the troops and providing social facilities. Involvement in the USO was one of the many ways in which the nation had come together to support the war effort, with nearly 1.5 million people having volunteered their services in some way.

The USO initially disbanded in 1947, but was revived in 1950 for the Korean War, after which it continued, also providing peacetime services. During the Vietnam War, USO social facilities ("USOs") were sometimes located in combat zones.

The USO has over 200 locations around the world in 14 countries (including the U.S.) and 27 states. During a gala marking the USO's 75th anniversary in 2016, retired Army Gen. George W. Casey Jr., the chairman of the USO Board of Governors, estimated that the USO has served more than 35 million Americans over its history.

Appendix

My Resignation Post 62

Gratitude and God Speed

To all the wise men and women who contributed recipes to this cookbook. Your recipes will remain in the revised Cookbook, which will be available for purchase for $20.00 at D&J Art Studio LLC, 3832 SE Dixie Highway, Stuart, FL 34997 and on amazon.com.

Home Going Celebration

Dr. Teresa, DC LMT and I rescued the Cookbooks from Post 62 and went

ALICIA HIGGINS
MY COMMISSION # HH 467714
EXPIRES: February 9, 2028

1

directly to American Legion Post 126, where Bruce Dennis Hudson's celebration of life was just beginning.

We found seats in the inside dining area downstairs, next to the food. We observed and listened as best we could. The military rituals took place in the upstairs dining room and the parking lot. Folks with accessibility issues stayed put and we could listen to folks tell stories about Bruce Hudson and his connections to the local community.

Aunt D is Food and Beverage

Aunt D told me the F&B rituals.

Cash bar, the bartender will run a tab. Food is all donated, some homemade, some from local caterers, so much that there will be leftovers for days.

The family goes through the food line first, then other folks can line up, grab plates and get their grub. Folks can go through the line as often as they want and there are take home boxes for folks who missed the service.

AMVETS 92 Rio Closes F&B

Debbie, Chief Cook and Bartender at AMVETS 92 Rio, greeted us as we arrived at Post 126 in Jensen. "Where's Chuck?" I asked. "He dropped me off and left." Said Debbie. She knew everyone there so she took off mingling.

Audrey, Chuck's live-in nurse (retired military RN), signed my application and I am a new member of the Auxiliary, AMVETS 92 in Rio.

Sunday Breakfast

Sunday breakfast is special at AMVETS 92 in Rio. The Post, usually packed, was quiet due to the rain.

Chuck called to ask if I was coming to breakfast. "I am just on my way out the door." Chuck ordered my breakfast and it arrived just as I walked in the door, along with strong black coffee.

Chuck helped me carry several copies of the offensive Cookbook, left on the table until we finished eating.

"Okay, what's up honey?"

"Would AMVETS Post 92 in Rio consider publishing this Cookbook? I will provide new ISBN and Library of Congress Registration Numbers so there is no connection with offensive Cookbook."

I showed Chuck the pages which Post 62 Auxiliary President Shirley Rearick found offensive and told him I could remove them.

"They stay." said Chuck.

I told Chuck that I wanted to dedicate the AMVETS 92 Cookbook to Bruce Dennis Hudson. Chuck nodded.

"I have to go take care of Audrey. I will review the Cookbook at home and talk to Mike."

Diane Freaney, Registered Democrat

What does it take to be bi-partisan? Does Shirley Rearick get to judge me?

3/11/24

Shirley played the "shame on me"; "shame on you", taking both sides.

In my resignation, the "shame game" is replayed.

"SHAME ON SHIRLEY," I am a Registered Democrat. It takes seconds to check my Voters Registration, or she could simply have asked me

"SHAME ON ME," for thinking I could work with a Leadership Team that values COMPETITION over COLLABORATION."

"War and Peace both require COLLABORATION. COMPETITION is deadly.

Post 62 Requires Books Destroyed

"And all current copies [of Post 62 Cookbooks] will be destroyed and not distributed." — Post 62 Auiliary President Shirley Rearick

The Nazi Book Burning of 1933

"On May 10, 1933, Nazis in Berlin burned works of Jewish authors, and the library of the Institut für Sexualwissenschaft, and other works considered "un-German."

Thank You Jesus!

The big guy upstairs is always on the job and lights the path forward.

What Shirley Rearick doesn't know and didn't ask, the offensive Cookbook has already been widely distributed to folks and organizations that have no connection to Post 62 in Stuart.

I am already getting requests to sign the offensive Cookbook(s) when I visit

AH

4

DF 3/11/24

their area. Hence, I honestly could not sign Shirley's book burning request.

The Big Guy Recommends

Mail copies of the AMVETS 62 Cookbook to folks and organizations that received the offensive cookbook and ask them to mail offensive cookbooks back.

[NOTE] These folks and organizations are NOT STUPID. They know that an original offensive Cookbook fetches much more money in book aftermarkets than $20.00.

Expunge | Erase | Obliterate the Offensive

We will absolutely expunge all mention of Post 62 from the books that we rescued from Post 62 on Saturday, February 17, 2024.

Our Team will notarized that document to insure compliance.

To all the folks, friends and collaborators at Post 62, I will miss you and the good times we had together.

Respectfully yours,

Diane M. Freaney
Martin Pearl Press LLC

ALICIA HIGGINS
MY COMMISSION # HH 467714
EXPIRES: February 9, 2028

5

DF 3/11/24

In Gratitude to God and Country

Ron Panton was my cousin Ronnie who I had not seen in 56 years. I am grateful to have seen him twice before he passed. Lynne and Ron met me at Alice's Family Restaurant in Stuart, where I now live, and once at their home in Boynton Beach. This is a photo on Ron and me sitting on the sofa in his living room, taken by his wife Lynne.

I was unable to attend the military funeral or the celebration back at the house. I hope to meet Ron's daughter, Lauren. Ron was so proud of Lauren and all his children and grandchildren.

Ronald M. Panton Obituary, by Lynne Panton

Ronald Panton of Boynton Beach, Florida, passed away on the 6th of May, 2021. He is survived his wife, Lynne Panton, his children William Panton, Dustin Panton, Lauren Panton, Michele Thompson, Christine King, Edward Dowd, and Donald Dowd; as well as his grandchildren, Gabriella and Jaxson Panton, Aimee Rob Reynard, Carrie Thompson, Chipper Thompson, Reeve King, CJ King, Brekka Dowd, Tori Dowd, Parker, Gavin, and Riley.

Ron Panton was born and raised in Boston, Massachusetts, to parents William and Pauline Panton. He was a very special man, as his daughter, Lauren, told me she was very blessed to have him. She said, not only was he the best Dad, he was my favorite person I've ever known. No one can replace that sweet man's soul. He is irreplaceable.

As far as being his wife, Ron gave me unconditional love. I am a better person to have had his love. I love him with all my heart, and I will miss him forever.

All Ron's children and grandchildren adored him. He was such a kind, loving man. Most of his friends said Ron was a cool guy! He loved football, the Patriots, and Tom Brady. He loved animals and always had at least four or five in the house. Ron also loved to ride Harleys, as he would say to me when he was going for a ride, "Honey - Just have to go and blow the dust off." He was a free spirit and the most important thing to him was his children and grandchildren.

Ron served in the Army and the Vietnam War. He worked for the United States Postal Service, and was a postman for almost 40 years. He loved everything about history, politics, and was an avid reader.

This brick has been placed in the Memorial Garden..

It's hard to describe Ron, he was one of a kind. The kindest, sweetest, loving man I have ever met. He said, "Please don't be sad when I'm gone. Live your life. Just remember me, and smile. Don't waste your time on worrying about things you can't control."

Up until the end, his positive outlook and sense of humor remained with him. In the last six months, we talked about eight hours a day. I could talk with him forever.

Ron, we will forever adore and remember you. We will stand taller and stronger knowing you are watching over us.

Ron will be buried in the South Florida National Cemetery. There will be a Celebration of Life for his family and friends.

Fly free, my love. You will be sorely missed, until we meet again!

To **Mariah Fenton Gladis**, Gestalt psychotherapist, who taught me never to stand tall and never give up. [https://mfgfoundation.org/]

To **Patti Messina**, Trauma (PSTD) Therapist, for helping me unlock my mind and put pen to paper as an Activist Author. [https://stuartcounseling.com/]

To my daughter A.E. Harter, grandson Duncan and all the folk I met along the way who have shared stories which informed my work.

Author's Bio

I have been seated at the table – rather, more accurately, been seated behind the white men seated at the table and told to hold my tongue – at the launch of some of the most radical new business models of the last century.

I have had a front row seat for every new finance and/or economic theory that came down the pike.

This is a dubious honor. I have watched these same new business models crash and burn, take jobs, destroy families, make ghost towns of cities, compromise our health and well-being, and rob us of our happiness. With the economic meltdown of 2008, I watched as my retirement fund plummeted 42% percent from the top of the market in 2007 to the bottom in 2009.

My life has been magical. I have over 70 years life/work experience and an excellent educational background.

Syracuse University (1965) – In my Accounting courses, I learned to fill in forms and play games with numbers. In Anthropology and Public Speaking, I learned storytelling and gained an appreciation of other cultures.

Harvard Business School – Corporate Finance Executive Education (1982) – I learned about OPM (Other People's Money), the strategy that brought the Global Financial Markets to their knees in 2008.

University of Pennsylvania – Organizational Dynamics (1999) – I learned that student's work is only valued when it follows a structured academic path. My Master's thesis – *A New Model for the Creative Use of College Endowments to Reduce College Tuition* would have prevented today's student loan crisis. Penn had no mechanism for "the administrators" to listen to students.

Bainbridge Graduate Institute (BGI) (2013) – I learned the importance of social media to listening deeply and delivering my message.

At an early age, I learned to communicate by listening. At my current age, I feel driven to share the knowledge and understanding amassed during my lifetime. Now I am speaking out.

—Diane Freaney, The Cat Lady

Coming Attractions

Complexity = Corruption Series

Complexity = Corruption / The US Mental Health System

Complexity = Corruption / The US Health System

Complexity = Corruption / The US Postal System

Complexity = Corruption / The US Insurance System

Complexity = Corruption / The US Emergency Management System

Complexity = Corruption / The US Federal Government System

Complexity = Corruption / The US Education System

Politics of Place Series

Living Life on Lake Osborne in John Prince Park

Higher Education Series

FAU and the Red Corvette Scandal